数据科学与大数据技术丛书

DATA SCIENCE
PRACTICE

数据科学实践

吕晓玲　李　舰　编著

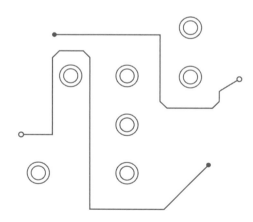

中国人民大学出版社
·北京·

数据科学与大数据技术丛书编委会

联合主任

蔡天文　　王晓军

委员（按姓氏拼音排序）

邓明华　　房祥忠　　冯兴东　　郭建华　　林华珍　　孟生旺
汪荣明　　王兆军　　张　波　　朱建平

前　言

21世纪人类社会步入了数据科学时代。随着现代社会的进步和通信技术的发展，在政治、经济、社会、文化等各个领域形成了规模巨大、增长与传递迅速、形式复杂多样、非结构化程度高的大数据。大数据的来源包括传感器、移动设备、在线交易、社交网络等，其形式可以是各种空间数据，报表统计数据，文字、声音、图像、超文本等各种环境和文化数据信息等。当下是一个海量数据广泛出现、运用逐渐普及的新的历史时期，新的社会环境需要我们认真研究与应对。

作为统计学科的教育工作者，我们要始终走在时代的前列，为学生开设最前沿的课程。数据科学实践这门课程是在数学、统计学的基础课以及机器学习、深度学习、分布式计算等专业课程之后开设的一门以实践为主的提升课程，目的是让学生对所学知识融会贯通，解决社会生产实践的具体问题。因此，本书的安排以案例教学为主，在第1章引言与第2章基础模型知识点介绍之后，第3、4、5章为三个大型机器学习案例，包括单机版实现以及Hadoop平台分布式实现。第6、7章为两个深度学习案例，需要用GPU服务器实现。读者可以从中国人民大学出版社网站（www.crup.com.cn）下载本书所有的原始数据和代码。

时代在发展，社会在进步。我们的教学工作也需要不断提升。本书仍有很多不足之处，希望读者不吝赐教，有机会再版的话，使其日臻完善。

目　录

第 1 章　引　言 ·· 1
  1.1　数据科学与人工智能时代 ················ 1
  1.2　数据智慧 ······································· 3
  1.3　本书内容安排 ································ 6

第 2 章　基础模型 ····································· 7
  2.1　机器学习方法 ································ 7
      2.1.1　双向聚类 ······························· 7
      2.1.2　基于邻居的推荐算法 ············ 12
      2.1.3　网络模型 ···························· 17
  2.2　深度学习 ······································ 19
      2.2.1　机器翻译模型 ······················ 19
      2.2.2　图像分析模型 ······················ 27

第 3 章　音乐风格识别 ···························· 34
  3.1　背景介绍 ······································ 34
  3.2　方法简介 ······································ 35
      3.2.1　音频数据和音频特征 ············ 35
      3.2.2　混合动力模型架构 ··············· 37
  3.3　描述分析 ······································ 39
      3.3.1　数据来源及简介 ··················· 39
      3.3.2　数据加工 ···························· 40
      3.3.3　音频特征提取 ······················ 42
  3.4　混合动力模型架构 ······················· 44
      3.4.1　两个基础模型的预测效果 ····· 44
      3.4.2　混合动力模型架构的预测效果 ··· 58
      3.4.3　工程优化 ···························· 62

第 4 章　航空数据案例分析 ···················· 66
  4.1　数据简介 ······································ 66
  4.2　单机实现 ······································ 69
      4.2.1　基于 Mysql 的数据预处理 ····· 69
      4.2.2　洛杉矶到波士顿航线的延误分析 ··· 72
      4.2.3　机场聚类分析 ······················ 82
      4.2.4　最短路径 ···························· 88
  4.3　分布式实现 ·································· 98
      4.3.1　基于 Hive 的数据预处理 ······· 98
      4.3.2　用 Spark 建立分类模型 ········ 101

第 5 章　公共自行车数据案例分析 ······ 106
  5.1　数据简介 ····································· 106
      5.1.1　交易流水表 ························· 106
      5.1.2　纽约市天气数据 ·················· 107
  5.2　单机实现 ····································· 108
      5.2.1　描述统计分析与可视化展现 ··· 108
      5.2.2　自行车角度的分析 ·············· 128
      5.2.3　单个站点借车量预测分析 ···· 137
  5.3　分布式实现 ································· 143
      5.3.1　数据预处理与描述统计 ······· 143

5.3.2 分布式预测模型 ·················· 145

第 6 章　机器翻译实例 ················· **151**
6.1 数据简介与数据预处理 ············ 151
  6.1.1 删除异常值 ·················· 151
  6.1.2 修改异常值及数据筛选 ······ 152
  6.1.3 BPE 分词 ····················· 160
6.2 数据描述统计 ························ 160
  6.2.1 句子长度统计 ················ 160
  6.2.2 词频统计 ······················ 160
  6.2.3 词性统计 ······················ 163
6.3 Seq2Seq+Attention 模型 ········· 167
  6.3.1 Seq2Seq 模型介绍 ··········· 167
  6.3.2 模型训练过程 ················ 169
  6.3.3 BLEU 值计算原理 ·········· 170
  6.3.4 模型训练结果 ················ 172

6.4 Transformer 模型 ··················· 173
  6.4.1 训练模型参数设置 ··········· 173
  6.4.2 训练结果 ······················ 173
6.5 模型对比 ······························ 174

第 7 章　眼底图像分析示例 ············ **175**
7.1 数据简介 ······························ 175
7.2 图像分割模型建立 ·················· 176
  7.2.1 数据预处理 ··················· 176
  7.2.2 模型训练 ······················ 180
7.3 基于图像的智能诊断 ··············· 181
  7.3.1 图像分割结果 ················ 181
  7.3.2 描述统计 ······················ 182
  7.3.3 诊断模型 ······················ 187

参考文献 ································· **190**

第 1 章

引 言

1.1 数据科学与人工智能时代

人类社会的进步永远伴随着科技的进步,科技的进步离不开数据分析. 数据挖掘 (Data Mining) 这一名词产生于 1990 年前后,之后迅速在学界和业界广泛应用与发展. 实际上,数据挖掘与统计数据分析的目标没有本质的差别. 按照《不列颠百科全书》,统计可以定义为收集、分析、展示、解释数据的科学. 这是历史相对悠久的统计在其发展过程中逐渐形成的被世人认可的定义. 它包含一系列概念、理论和方法,有比较稳定的知识结构和体系. 数据挖掘也完全符合这一定义,但它的历史较短,初期主要由计算机科学家开创,脱离了传统统计的体系,因此有其自身的特点. 数据挖掘有时也称数据库的知识发现 (Knowledge Discovery in Databases, KDD). 严格来讲这两个概念并不完全一致. 目前使用更多的术语是机器学习 (Machine Learning). 从统计学的角度则称统计机器学习 (Statistical Machine Learning) 或统计学习 (Statistical Learning).

一般认为,麦肯锡公司的研究部门——麦肯锡全球研究院 (MGI) 在 2011 年首先提出大数据时代 (Age of Big Data) 的概念,这一概念的横空出世在全球引起广泛反响. 早在 2001 年,美国信息咨询公司 Gartner 的分析师 Doug Laney 就从数据量 (Volume)、多样化 (Variety) 和快速化 (Velocity) 三个维度分析了在数据量不断增长的过程中数据分析所面临的挑战和机遇. 在大数据这一概念被广泛传播后,IBM 副总裁 Steven Mills 于 2011 年在此基础上提出大数据的第四个维度——价值密度 (Veracity). 人们普遍认为大数据蕴含巨大的价值,而如何从中快速准确地提取真实有价值的信息是大数据处理技术的关键.

大数据,是指随着现代社会的进步和通信技术的发展,在政治、经济、社会、文化各个领域形成的规模巨大、增长与传递迅速、形式复杂多样、非结构化程度高的数据或者数据集. 它的来源包括传感器、移动设备、在线交易、社交网络等,其形式可以是各种空间数据,报表统计数据,文字、声音、图像、超文本等各种环境和文化数据信息等. 大数据时代是一个海量数据开始广泛出现、海量数据的运用逐渐普及的新的历史时期,也形成了需要我们认真研究与应对的新的社会环境. 数据科学 (Data Science) 一词应运而生. 它可以看作数学逻辑和统计批判性思维、计算机科学以及实际领域知识这三者的交集 (见图 1.1).

人脑是由大量神经细胞相互连接形成的一种复杂的信息处理系统,长期的自然进化使人脑具备很多良好的功能,如分布式表示和计算、巨量并行性、学习能力、推广能力、容错能力、自适应性等. 人工神经网络 (Artificial Neural Networks, ANN),简称神经网络 (Neural Networks, NN),是通过对人脑神经系统的抽象和建模得到的简化模型,是一种具有大量连接的并行分布式处理器,由简单的处理单元组成,具有通过学习来获取知识并解决问题的能力.

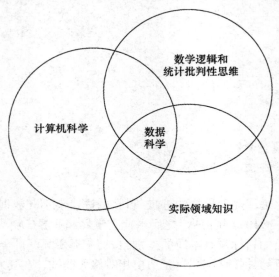

图 1.1　数据科学

人工神经网络已有 70 多年的研究历史,其发展过程曲折,几经兴衰. 人工神经网络研究的先驱为生理学家 McCulloch 和数学家 Pitts,他们于 1943 年在神经细胞生物学基础上,从信息处理的角度出发提出形式神经元的数学模型 (McCulloch and Pitts, 1943),开启了人工神经网络研究的第一个热潮. 然而, 1969 年人工智能创始人之一的 Minsky 和计算机科学家 Papert 在《感知器》一书中指出感知器模型的缺陷,由此引发了人工神经网络发展史上长达十几年的低潮期. 1982 年, 美国物理学家 Hopfield 提出了一种新颖的 Hopfield 网络模型, 这标志着人工神经网络研究工作的复苏. 随后以 Rumelhart 和 McClelland 为代表的科学家小组于 1986 年出版了《并行分布式处理》一书的前两卷, 该书介绍了并行分布式处理网络思想, 发展了适用于多层神经网络模型的反向传播算法, 克服了感知器模型继续发展的重要障碍, 由此引发了神经网络研究的第二个热潮. 然而从 90 年代开始, 人工神经网络逐渐受到冷落. 这某种程度上是由于以支持向量机和组合算法为代表的统计学习的兴起, 但更重要的是神经网络的巨大计算量和优化求解难度使其只能包含少量隐层, 从而限制了其在实际应用中的性能. 2006 年, 多伦多大学计算机系教授 Geoffrey Hinton 及其学生 Salakhutdinov 在《科学》(Science) 上发表文章, 认为多隐层的人工神经网络具有优异的特征学习能力, 而对于多隐层神经网络在训练上的困难, 可以通过 "逐层初始化" 来有效克服, 由此 Hinton 等人进一步提出了深度学习的概念, 开启了深度学习的研究浪潮. 目前, 深度学习引起学界与业界的广泛关注, 在语音识别、图像识别、自然语言处理等领域获得了突破性进展. 在深度学习基础上发展的强化学习, 则在人机博弈、AI 游戏等领域取得了惊人的成就, 唤醒了人工智能时代.

1.2 数据智慧

2016 年第 1 期《中国计算机学会通讯》刊登了美国加利福尼亚大学伯克利分校统计系郁彬教授 (美国科学院院士、美国艺术与科学学院院士) 的一篇文章:《数据科学中的数据智慧》, 英文原文的网址链接是 http://www.odbms.org/2015/04/data-wisdom-for-data-science/.

郁彬教授深入地讨论了应用统计方法解决实际问题应该注意的事项, 明确提出 "数据智慧" 是应用统计学概念的核心. 希望读者认真阅读这篇文章并思考: 在大数据时代, 统计数据分析工作者的任务和使命是什么? 我们怎样才能正确应用统计方法解决实际问题?

在大数据时代, 学界和业界的大量研究都是关于如何以一种可扩展和高效率的方式对数据进行存储、交换和计算 (通过统计方法和算法). 这些研究非常重要. 然而, 只有对数据智慧 (Data Wisdom) 给予同等程度的重视, 大数据 (或者小数据) 才能转化为真正有用的知识和可采纳的信息. 换言之, 我们要充分认识到, 只有拥有足够数量的数据, 才有可能对复杂度较高的问题给出较可靠的答案. 数据智慧对于我们从数据中提取有效信息和确保没有误用或夸大原始数据是至关重要的.

"数据智慧" 一词是对应用统计学核心部分的重新定义. 这些核心部分在伟大的统计学家 (或者说数据科学家) John W. Tukey (1962) 的文章和 George Box (1976) 的文章中都有详细介绍.

将统计学核心部分重新命名为 "数据智慧" 非常必要, 因为它比 "应用统计学" 这个术语起到了更好的概括作用. 这一点最好能让统计学领域之外的人也了解到. 这样一个有信息量的名称可以使人们意识到应用统计作为数据科学的一部分的重要性.

数据智慧是将领域知识、数学和方法论与经验、理解、常识、洞察力以及良好的判断力相结合, 思辨性地理解数据并依据数据做决策的一种能力.

数据智慧是数学、自然科学和人文主义三方面能力的融合, 是科学和艺术的结合. 如果没有实践经验者的指导, 仅靠读书很难掌握数据智慧. 学习它的最好方法就是和拥有它的人一起共事. 本书通过问答的方式帮助你形成和培养数据智慧. 下面有 10 个方面的基本问题, 我鼓励人们在开始从事数据分析项目时或者在项目进行过程中经常问问自己这些问题. 这些问题是按照一定顺序排列的, 但是在不断重复的数据分析过程中, 这个顺序完全可以打乱.

这些问题也许无法详尽彻底地解释数据智慧, 但是它们体现了数据智慧的一些特点.

1. 要回答的问题

数据科学问题最初往往来自统计学或者数据科学以外的学科. 例如, 神经科学中的一个问题: 大脑是如何工作的? 或银行业中的一个问题: 该向哪组顾客推广新服务? 要解决这些问题, 统计学家必须与这些领域的专家合作. 这些专家会提供有助于解决问题的领域知识、早期的研究成果、更广阔的视角, 甚至可能对问题进行重新定义. 而与这些专家 (他们往往很忙) 建立联系需要很高超的人际交流技巧.

与领域专家的交流对于数据科学项目的成功是必不可少的. 在数据来源充足的情况下, 常见的情形是在收集数据前还没有精确定义要回答的问题. 我们发现自己处在图基所说的"探索性数据分析"(Exploratory Data Analysis, EDA) 的游戏中. 我们寻找需要回答的问题, 然后不断地重复统计调查过程 (就像 George Box 的文章中所述). 由于误差的存在, 我们谨慎地避免对数据中出现的模式进行过拟合. 例如, 当同一份数据既用于对问题进行建模又用于对问题进行验证时, 就会发生过拟合. 避免过拟合的黄金准则就是对数据进行分割, 在分割时考虑到数据潜在的结构 (如相关性、聚类性、异质性), 使分割后的每部分数据都能代表原始数据. 其中一部分用来探索问题, 另一部分用于通过预测或者建模来回答问题.

2. 数据收集

什么样的数据与第一个方面要回答的问题最相关?

实验设计 (统计学的分支) 和主动学习 (机器学习的分支) 中的方法有助于回答这个问题. 即使在数据收集好了以后考虑这个问题也是很有必要的. 因为对理想的数据收集机制的理解可以暴露出实际数据收集过程的缺陷, 能够指导下一步分析的方向.

下面的问题会对提问有所帮助: 数据是如何收集的? 在哪些地点? 在什么时间段? 是谁收集的? 用什么设备收集? 中途更换过操作人员和设备吗? 总之, 试着想象自己在数据收集现场.

3. 数据含义

数据中的某个数值代表什么含义? 它测量了什么? 它是否测量了需要测量的? 哪些环节可能会出错? 在哪些统计假设下可以认为数据收集没有问题?(对数据收集过程的详细了解在这里会很有帮助.)

4. 相关性

收集来的数据能够完全或部分回答要研究的问题吗? 如果不能, 还需要收集其他什么数据? 第二个方面的问题的要点在此处同样适用.

5. 问题转化

如何将第一个方面的问题转化成一个与数据相关的统计问题, 使之能够很好地回答原始问题? 有多种转换方式吗? 比如, 我们可以把问题转换成一个与统计模型有关的预测问题或者统计推断问题吗? 在选择模型前, 请列出与回答实质性问题相关的每一种转化方式的优点和缺点.

6. 可比性

各数据单元是不是可比的, 或经过标准化处理后可视为可交换的? 苹果和橘子是否被组合在一起? 数据单元是不是相互独立的? 两列数据是不是同一个变量的副本?

7. 可视化

观察数据 (或其子集), 制作一维或二维图表, 并检验这些数据的统计量. 询问数据范围是什么、数据是否正常、是否有缺失值, 使用多种颜色和动态图来表明这些问题. 是否有意料之外的情况? 值得注意的是, 我们大脑皮层的 30% 是用来处理图像的, 所以可视化方法在挖掘数据模式和遇到特殊情况时非常有效. 通常情况下, 为了找到大数据的模式, 在某些模型建立之后使用可视化方法最有用, 比如计算残差并进行可视化展示.

8. 随机性

统计推断的概念 (比如 p 值和置信区间) 都依赖于随机性. 数据中的随机性的含义是什么? 我们要使统计模型的随机性尽可能明确. 哪些领域知识支持统计模型中的随机性描述? 一个表现统计模型中随机性的最好例子是因果关系分析中内曼 – 鲁宾 (Neyman-Rubin) 的随机分组原理 (在 AB 检验中也会使用).

9. 稳定性

你会使用哪些现有的方法? 不同的方法会得出同一个定性的结论吗? 举个例子, 如果数据单元是可交换的, 可以通过添加噪声或二次抽样对数据进行随机扰动 (一般来说, 应确定二次抽样样本遵守原样本的底层结构, 如相关性、聚类特性和异质性, 这样二次抽样样本才能较好地代表原始数据), 这样做得出的结论依然成立吗? 我们只相信那些能通过稳定性检验的方法。稳定性检验简单易行, 能够抗过拟合和过多假阳性的发现, 具有可重复性 (要了解关于稳定性重要程度的更多信息, 请参见文章 "*Stability*" (Yu Bin, 2013)).

可重复性研究最近在学术界引起很多关注 (请参见《自然》(*Nature*) 特刊, http://www.nature.com/news/reproducibility-1.17552)).《科学》的主编 Marcia McNutt 指出, "实验再现是科学家用以增加结论信度的一种重要方法". 同样, 商业和政府实体也应该要求从数据分析中得出的结论在用新的同质数据检验时是可重复的.

10. 结果验证

如何知道数据分析做得好不好呢? 衡量标准是什么? 可以考虑用其他类型的数据或者先验知识来验证, 不过可能需要收集新的数据.

在数据分析时还有许多其他问题要考虑, 但希望上面这些问题能使你对如何获取数据智慧产生一点感觉. 这些问题的答案需要在统计学之外获得. 要找到可靠的答案, 有效的信息源包括 "死的" (如科学文献、报告、书籍) 和 "活的"(如人). 出色的人际交流技能使寻找正确信息源的过程简单许多，即使在寻求"死的"信息源的过程中也是这样．因此，为了获取充足的有用信息，人际交流技能十分重要，知识渊博的人通常是你最好的指路人．

1.3 本书内容安排

本书第 2 章介绍书中案例所需的统计方法．第 3～5 章为三个大型机器学习案例，包括单机版实现以及 Hadoop 平台分布式实现．第 6、7 章为两个深度学习案例，需要 GPU 服务器实现．

本书案例所使用的主要环境为 Python3.6+Tensorflow1.14．还有一些案例的图表展示使用到了 R3.5.0．

第 2 章 基础模型

本章简要介绍本书各案例所使用的模型. 相对基础的机器学习模型, 比如 K 均值聚类、时间序列模型、logisitic 回归、随机森林等, 在此略过, 不做介绍. 我们只介绍一些相对复杂的方法, 以及深度学习的相关概念和模型.

2.1 机器学习方法

2.1.1 双向聚类

1. 双向聚类概述

一般的聚类方法是根据变量的取值对观测进行聚类. 双向聚类则是同时考虑观测与变量的差异.

在过去的十余年时间里, 双向聚类在双向数据分析 (Two-Way Data Analysis) 领域越来越受欢迎, 在基因数据分析与商业活动等领域有着广泛的应用.

所谓基因表达数据, 就是生物学上通过某种手段测定的表征基因表达强度的数据. 这些数据一般都存放在一个基因表达矩阵中, 矩阵的每一行代表一个基因, 每一列代表一个条件, 每个元素代表对应行的基因在对应列的条件下表达出来的强度.

得到一个基因表达矩阵以后, 对基因 (行) 或者条件 (列) 进行聚类是我们通常要做的事情. 比如, 可以考虑诸基因在不同条件下的表达情况, 使用 K 均值方法对基因进行聚类; 或者反之, 考虑诸条件下不同基因的表达情况, 使用 K 均值方法对条件进行聚类.

遗憾的是, 像 K 均值聚类这样的传统的聚类方法在基因表达数据上并不总是能够工作得很好, 它们往往会遗漏掉一些有意思的模式. 这是因为传统的聚类方法一般是根据基因在所有条件下的表达情况对其进行聚类的, 因此, 这些方法只能发现某种全局模式. 然而, 有些基因只是在某些特定的条件下才表达, 在其余的条件下是不表达的. 比如, A 基因和 B 基因在 C_1 条件下是协同表达的, 因而可以在 C_1 条件下归为一类, 但是它们在 C_2 条件下却并没有什么协同联系; 与此同时, A 基因和 D 基因在 C_2 条件下是协同表达的, 因而可以视为一类, 但是在 C_1 条件下却不能视为一类. 这告诉我们, 部分基因在特定条件下才能聚为一类, 在全部条件下考察的话, 很可能错失这种有意义的局部模式.

另外，传统的聚类方法中，类与类之间是相互排斥的，一般不允许类之间有重叠. 可是在现实生活中，同一个基因参与不同的细胞过程是很平常的事情，因此该基因理所当然可以在不同的条件下被划分到不同的类别中，这在上面的例子中有所体现. 所以，我们需要一种新的聚类方法来照顾到基因数据中出现的这些局部模式和可重叠性.

双向聚类就是这样一种方法，它在聚类的过程中综合考虑基因和条件，试图发现一些让人感兴趣的局部类. 在更一般的框架下，给定 n 行 m 列的实值矩阵 $A = (X,Y)$，其中 X 和 Y 分别为 A 的行指标和列指标集合. 我们想要找到一个子矩阵 $B_k = (I_k, J_k)$，其中 $I_k \subseteq X, J_k \subseteq Y$，使得 B_k 能够具有某种意义上的同质性，就像传统聚类中类内的元素具有某种意义上的同质性那样. 这样的子矩阵称为双向类 (Bicluster)，这种寻找双向类的聚类方法叫做双向聚类 (Biclustering).

表 2.1 是一个双向聚类在商业领域的例子，每一行代表一个用户，每一列代表一个产品. 每个数据点代表用户购买了该产品，购买为 1, 不购买为 0; 也可以是用户对产品的评分 (连续型数据). 我们可以根据用户在某些产品 (而不是全部产品) 上的喜好进行用户和产品的双向聚类.

表 2.1 用户和产品的数据

用户	产品			
	V_1	V_2	$\cdots$	V_m
U_1	1	0	$\cdots$	1
U_2	0	1	$\cdots$	1
$\vdots$	$\vdots$	$\vdots$		$\vdots$
U_n	1	0	$\cdots$	0

双向聚类还可以用于文本挖掘以及其他拥有类似结构数据的领域. 比如，文本挖掘中经常会碰到这样的矩阵，它的每一行代表一个文档，每一列代表一个单词，每个元素则表示对应列的单词在对应行的文档中出现的频率. 对这样的文档 – 词频矩阵，也可以使用双向聚类来发现我们感兴趣的局部模式.

同质性因方法而异，不同的双向聚类算法一般定义不同的同质性指标. 比如，同质可以指一个子矩阵中包含完全相同的或者近似完全相同的元素，也可以指一个子矩阵的每一行或每一列都包含相同的元素，还可以指一个子矩阵中的元素随着行指标和列指标的增长呈现递进的趋势. 当我们定义了一种新的合理的同质性和双向类，并设计了一个有效的算法来找出这些双向类时，一种新的双向聚类方法也就建立起来.

> **思考:**
> - 双向聚类还适用于哪些数据结构?

2. BIMAX 算法

双向聚类有很多算法，大部分采用迭代方法，即在发现 $n-1$ 个双向类的情况下，发现下一个双向类. 这里介绍 BIMAX 算法.

BIMAX (Binary Inclusion-Maximal Biclustering Algorithm) 算法是由 Prelic et al. (2006) 提出的. 如表 2.1 所示, 若矩阵的每个元素只有两个可能的值 (0 或者 1), 原始数据可以表示为 n 行 m 列的二分数据矩阵 $\boldsymbol{E}^{n\times m}$. 一个双向类 (G,C) 对应于列集合 $C \subseteq \{1,2,\cdots,m\}$ 和行集合 $G \subseteq \{1,2,\cdots,n\}$. 也就是说, (G,C) 定义了所有元素为 1 的一个子矩阵.

在该定义下, 每个值为 1 的元素 e_{ij} 本身就代表一个双向类, 但 BIMAX 寻找的是最大包含 (inclusion-maximal) 的类, 即对任意一个类 $(G,C) = \{i \in G, j \in C : e_{ij} = 1\}$, 不存在另一个类 (G',C'), 使得 $(G',C') \subset (G,C), (G',C') \neq (G,C)$.

我们采用如下 BIMAX 的迭代算法:
- 第一步: 重排行和列, 使得 1 集中在矩阵的右上角.
- 第二步: 将矩阵分为两个子矩阵, 若一个子矩阵中只有 1, 则返回该子矩阵.

为了得到一个令人满意的结果, 该方法需要从不同的起点重复几次. BIMAX 尝试识别出 $\boldsymbol{E}$ 中只包含 0 的区域, 可以在进一步的分析中排除这些区域. 因此当 $\boldsymbol{E}$ 为稀疏矩阵时, BIMAX 具有独特的优势. 此外, BIMAX 需要的存储空间和运算时间都较少.

3. CC 算法

Cheng 和 Church (2000) 提出一种双向聚类算法, 为方便起见, 我们把他们提出的算法称为 CC 算法.

对于一个矩阵 $\boldsymbol{A} = (I, J) = (a_{ij})$, 其中, I 和 J 分别为行指标集和列指标集, 记

$$a_{iJ} = \frac{1}{|J|} \sum_{j \in J} a_{ij}$$

$$a_{Ij} = \frac{1}{|I|} \sum_{i \in I} a_{ij}$$

$$a_{IJ} = \frac{1}{|I||J|} \sum_{i \in I, j \in J} a_{ij}$$

分别为行均值、列均值和矩阵均值. 为了给矩阵 $\boldsymbol{A}$ 一个同质性的度量, CC 算法定义了得分函数:

$$H(I,J) = \frac{1}{|I||J|} \sum_{i \in I, j \in J} (a_{ij} - a_{iJ} - a_{Ij} + a_{IJ})^2$$

得分函数 $H(I,J)$ 刻画了矩阵 I,J 的波动程度. 一个矩阵的得分越低, 说明该矩阵的同质性越高. 在上述同质性指标下, CC 算法进一步定义了 δ-双向类: 若存在一个 $\delta > 0$, 使得矩阵 I,J 的 $H(I,J) \leqslant \delta$, 则称 I,J 为 δ-双向类. CC 算法的目标就是在给定初始矩阵 $\boldsymbol{A}$ 和阈值 δ 的条件下, 尽可能找到尺寸比较大的 δ-双向类.

初始矩阵 $\boldsymbol{A} = (I,J)$ 的得分 $H(I,J)$ 一般都要比 δ 大, 因此, 我们希望通过不断删除一些行或者列, 使矩阵的得分持续降低, 直到比 δ 小. 哪些行或者列被删除后能够降低一个矩阵的得分呢? 为此, 我们需要对每一行和每一列的波动程度进行刻画.

$$d(i) = \frac{1}{|J|} \sum_{j \in J} (a_{ij} - a_{iJ} - a_{Ij} + a_{IJ})^2 \tag{2.1.1}$$

$$d(j) = \frac{1}{|I|} \sum_{i \in I} (a_{ij} - a_{iJ} - a_{Ij} + a_{IJ})^2 \qquad (2.1.2)$$

式 (2.1.1) 刻画了第 i 行的波动程度, 式 (2.1.2) 刻画了第 j 列的波动程度. Cheng 和 Church 证明了, 如果删除那些 $d(i) > H(I, J)$ 的行, 那么得到的新矩阵的得分一定比 $H(I, J)$ 小; 如果删除那些 $d(j) > H(I, J)$ 的列, 那么得到的新矩阵的得分也一定比 $H(I, J)$ 小. 这就指导我们提出如下单节点删除算法.

算法 1: 单节点删除

输入: 原始矩阵 A, 行指标集 I, 列指标集 J, 阈值 $\delta > 0$

输出: 行指标集 $I' \subseteq I$, 列指标集 $J' \subseteq J$, 使得 $H(I', J') \leqslant \delta$

1 计算诸如 a_{ij}, a_{iJ}, a_{Ij} 和 $H(I, J)$
2 **while** $H(I, J) > \delta$ **do**
3   找到行指标 $r = \text{argmax} d(i)$ 和列指标 $c = \text{argmax} d(j)$
4   如果 $d(r) > d(c)$, 就将第 r 行删除, 否则就将第 c 列删除
5   更新行指标集 I 和列指标集 J, 重新计算诸如 a_{ij}, a_{iJ}, a_{Ij} 和 $H(I, J)$
6 **end**
7 **return** 最终的行列指标集

理论上, 上述单节点删除算法最后一定会返回一个 δ-双向类, 但是由于在每次循环中只能删除一行或者一列, 所以该算法的速度比较慢. 因此, 人们提出多节点删除算法来加速这一过程.

算法 2: 多节点删除

输入: 原始矩阵 A, 行指标集 I, 列指标集 J, 阈值 $\delta > 0$, 调整因子 $\alpha > 1$

输出: 行指标集 $I' \subseteq I$, 列指标集 $J' \subseteq J$, 使得 $H(I', J') \leqslant \delta$

1 计算诸如 a_{ij}, a_{iJ}, a_{Ij} 和 $H(I, J)$
2 **while** $H(I, J) > \delta$ **do**
3   删除所有 $d(i) > \alpha H(I, J)$ 的行
4   更新行指标集 I, 重新计算诸如 a_{ij}, a_{iJ}, a_{Ij} 和 $H(I, J)$
5   删除所有 $d(j) > \alpha H(I, J)$ 的行
6   更新列指标集 J, 重新计算诸如 a_{ij}, a_{iJ}, a_{Ij} 和 $H(I, J)$
7   如果行列指标集都不再变化, 那么退出循环
8 **end**
9 **return** 转到单节点删除算法

关于多节点删除算法, 有几点需要注意. 首先, 每次循环中会有较多的行和列被删除, 为了避免一次删除过多, 我们在单节点删除算法的基础上加上一个调整因子 $\alpha > 1$, 只有波动程度比 $\alpha H(I, J)$ 还要大的那些行与列才能删除. 其次, 多节点删除算法一般不单独使用, 而是配合单节点删除算法使用, 我们先使用多节点删除算法将矩阵的尺寸降下来, 然后使用单节点删除算法对其进行修剪.

通过两次删除算法得到的子矩阵一定是 δ-双向类, 但未必是最大的 δ-双向类, 因此 CC

算法还有一个节点添加的过程, 将那些 $d(j) \leqslant H(I, J)$ 的列和 $d(i) \leqslant H(I, J)$ 的行添加进去.

算法 3: 节点添加

输入: δ-双向类 $\boldsymbol{A} = (I, J)$
输出: 行指标集 $I' \subseteq I$, 列指标集 $J' \subseteq J$, 使得 $H(I', J') \leqslant \delta$
1　计算诸如 a_{ij}, a_{iJ}, a_{Ij} 和 $H(I, J)$
2　**while** $H(I, J) > \delta$ **do**
3　　将 $j \notin J$ 且 $d(j) \leqslant H(I, J)$ 的那些列添加进去
4　　更新列指标集 J, 重新计算诸如 a_{ij}, a_{iJ}, a_{Ij} 和 $H(I, J)$
5　　将 $i \notin I$ 且 $d(i) \leqslant H(I, J)$ 的那些行添加进去
6　　更新行指标集 I, 重新计算诸如 a_{ij}, a_{iJ}, a_{Ij} 和 $H(I, J)$
7　　如果行列指标集都不再变化, 那么退出循环
8　**end**
9　**return** 最终的行列指标集

Cheng 和 Church 已经证明了, 上述节点添加的过程不会造成矩阵得分 $H(I, J)$ 增大, 因而最终可以得到尽可能大的 δ-双向类.

这样, 通过多节点删除、单节点删除、节点添加这三步, 就能够在原始矩阵中找到一个 δ-双向类. 为了继续寻找其他可能的双向类, 要用均匀分布随机数覆盖上一步找到的双向类, 然后再重复上述三步.

完整的 CC 算法如下:

算法 4: CC 算法

输入: 原始矩阵 $\boldsymbol{A}$, 双向类的数目 N, 阈值 $\delta > 0$, 调整因子 $\alpha > 1$
输出: N 个双向类
1　计算诸如 a_{ij}, a_{iJ}, a_{Ij} 和 $H(I, J)$
2　**for** $i = 1, 2, \ldots, N$ **do**
3　　多节点删除
4　　单节点删除
5　　节点添加
6　　将找到的双向类输出
7　　将原始矩阵对应于该双向类的那部分元素用均匀分布随机数进行覆盖
8　**end**

除了本书介绍的 BIMAX 算法和 CC 算法, 还有一些其他的双向聚类方法, 读者可以自行阅读相关文献.

> **思考**:
> - CC 算法相较于 BIMAX 算法的优势在哪里?
> - CC 算法中进行节点添加后有必要再进行节点删除吗? 为什么?

2.1.2 基于邻居的推荐算法

推荐系统是在信息过载时信息拥有者向它的受众进行有选择的推送的系统. 比如, 当你打开平时上网看电影的界面时, 服务器会从数据库中调取你的观看记录和喜好, 给你推荐电影. 有时候网站给你推荐与以往看过的题材类似的电影、相同演员或导演的电影、不常见的小众电影等, 这个过程往往发生在刚刚看完一部电影之后. 有时候你在网络购物平台下了订单之后, 浏览器的侧边会出现类似物品的广告, 这些内容随着购物的不同而变化, 这就是推荐系统的算法在起作用. 现在很多通过手机发送的广告会随着地理位置不同而呈现不同的内容, 并且与用户的购物行为、喜好等高度相关, 具有个性化的特点. 随着精准营销的兴起, 推荐系统在越来越多的电商平台、移动互联网、基于位置的服务中扮演越来越重要的角色, 将其形容为核心算法一点也不为过. 将正确的商品 (或服务) 在正确的时间、正确的地点推荐给正确的人, 这在商业上是有巨大价值的. 准确的推荐能够给用户的生活带来很大的便利.

前面提到的几个场景下的推荐基本上对应着传统推荐系统的几个设计出发点: 基于物品或用户的相似度的推荐、基于潜在因子 (或称隐因子) 的推荐等. 简单地说, 基于用户的推荐就是针对每个用户, 寻找与他的喜好相似的其他用户, 并将相似用户的物品推荐给该用户; 基于物品的推荐则是寻找相似的物品, 向每位用户推荐与他喜欢的物品相似的物品. 这里的"喜欢"在电影推荐的场景下可以是用户是否浏览、是否关注的行为; 在网购的场景下可以是用户点击具体商品的行为, 在两个场景下也可以是对电影或商品的打分或评价行为. 本节介绍这部分内容. 其他的推荐算法还包括基于潜在因子与矩阵分解的方法和基于深度学习的方法, 有兴趣的读者可自行查阅相关资料.

1. 基于邻居的预测算法

总体来看, 基于用户或者物品的推荐属于基于邻居的推荐方法. 所谓邻居, 是指与一个对象比较近的其他对象. 所谓物以类聚、人以群分, 距离比较近的对象往往具有相近的特征. 比如, 在 K 近邻 (KNN) 算法当中, 我们认为与一个点欧氏距离 (或者以其他方式定义的距离) 最近的 K 个点是这个点的邻居 (注意在计算距离时只利用协变量), 并认为这个点的目标变量的类别 (分类问题) 或数值 (回归问题) 可以通过所有邻居的类别的众数 (或数值的平均) 来预测. 基于用户或物品相似性的推荐利用的是类似的思想.

(1) 连续型评分

以对电影进行评分为例, 设共有 N 个用户, M 部电影, 评分矩阵记为 $\boldsymbol{R}_{N \times M}$, r_{ui} 表示第 u 个用户对第 i 部电影的评分. 注意在评分矩阵当中存在很多缺失值, 表示用户并未观看某些电影或者未对某些电影进行评分. 先从基于用户的角度来看, 如果用户 u 对电影 i 尚未评分, 记 $N(u)$ 为用户 u 的邻居, $N_i(u)$ 为所有评价过电影 i 的用户中 u 的邻居, 假定这里定义的"邻居"可以反映出喜好或者评分上的相似性, 那么可以利用评价过电影 i 的 u 的邻居的评分平均值来预测用户 u 对电影 i 的评分:

$$\hat{r}_{ui} = \frac{1}{|N_i(u)|} \sum_{v \in N_i(u)} r_{vi}$$

如果对于不同的邻居, 用户 u 与他们之间有不同的相似度, 用 ω_{uv} 表示用户 u, v 之间相

似度的大小, 则可以利用加权平均来进行预测:

$$\hat{r}_{ui} = \frac{\sum\limits_{v \in N_i(u)} \omega_{uv} r_{vi}}{\sum\limits_{v \in N_i(u)} |\omega_{uv}|}$$

式中, 相似度 ω_{uv} 可以大于 0, 表示用户 u 与用户 v 的喜好正向相似; 也可以小于 0, 表示用户 u 与用户 v 之间的喜好相反. 这些邻居都能对预测用户 u 的评分起到显著的作用, 所以上式分子中的 ω_{uv} 不需要取绝对值, 分母则需要进行取绝对值的运算.

有时候不同的人对相同程度的 "认可" 打分差异很大, 如在以 100 分为基准的评价下, 有些人习惯在 $60 \sim 80$ 之间打分, 有些人习惯在 $40 \sim 70$ 之间打分, 所以需要对不同的人的分数进行标准化处理. 设 h 是一个标准化函数, 可将不同用户对不同物品的打分 r_{ui} 映射到某一指定的区间 $[a,b]$ 上. 在此基础上计算 r_{ui} 的预测值, 再通过 h 的反函数将其映射回原始的取值范围 (Ricci et al., 2011). 具体公式如下:

$$\hat{r}_{ui} = h^{-1} \left(\frac{\sum\limits_{v \in N_i(u)} \omega_{uv} h(r_{vi})}{\sum\limits_{v \in N_i(u)} |\omega_{uv}|} \right)$$

从基于物品的角度来看, 与基于用户相似度的评分预测类似, 记 $N(i)$ 为物品 i 的 "邻居", 即最像物品 i 的物品集合, $N_u(i)$ 表示用户 u 评分过的物品中最像物品 i 的物品集合, 则用户 u 对物品 i 的评分预测为:

$$\hat{u}_{ui} = \frac{\sum\limits_{j \in N_u(i)} \omega_{ij} h(r_{uj})}{\sum\limits_{j \in N_u(i)} |\omega_{ij}|}$$

式中, ω_{ij} 表示物品 i,j 之间相似度的权重.

当需要调整评分尺度时, 可以引入标准化函数 h 进行预测:

$$\hat{u}_{ui} = h^{-1} \left(\frac{\sum\limits_{j \in N_u(i)} \omega_{ij} h(r_{uj})}{\sum\limits_{j \in N_u(i)} |\omega_{ij}|} \right)$$

(2) 离散类别评分

除了上述连续型评分, 往往还有离散类别的评分, 比如 "好" "中" "差" 用数字 1,2,3 表示. 以基于用户相似度的分类为例, 假设一部电影的评分可以从 $S = 1, 2, \cdots, K$ 共 K 个分数选项中进行选择, 尝试利用邻居的打分来确定用户 u 最可能的打分. 我们利用邻居在第 r 级分数上的打分情况来预测用户 u 打 r 分的可能性 $(r = 1, 2, \cdots, K)$.

$$\hat{u}_{ir} = \sum\limits_{v \in N_i(u)} \delta(r_{vi} = r) \omega_{uv}$$

式中, $\delta(\cdot)$ 为示性函数, 取值为 1 或 0; ω_{uv} 为用户 u 与 v 的相似度.

对所有的 K 个分数分别计算评分 $\hat{u}_{ir}$ 后, 用最大的 $\hat{u}_{ir}$ 对应的 r 作为预测的评分:

$$\hat{r}_{ir} = \arg\max_{r} \hat{u}_{ir}$$

这种方法同样会受到人群和物品的打分区域的差异的影响, 引入标准化函数 h 的版本如下:

$$\hat{r}_{ui} = h^{-1}\left(\arg\max_{r \in S'} \sum_{v \in N_i(u)} \delta(h(r_{vi}) = r)\omega_{uv}\right)$$

式中, S' 为前面的评分值集合 S 对应的标准化后的评分值集合.

同理, 对于基于物品相似度的分类问题, 利用以下模型进行预测:

$$\hat{r}_{ui} = h^{-1}\left(\arg\max_{r \in S'} \sum_{j \in N_u(i)} \delta(h(r_{uj}) = r)\omega_{ij}\right)$$

2. 基于邻居的预测的三要素

上文介绍了基于邻居的预测算法的原理, 实际使用这一算法的过程中, 有三个基本要素需要进行考察, 这三个基本要素为: 邻居选择、相似度计算和评分标准化, 下面一一介绍 (Ricci et al., 2011).

(1) 邻居选择

邻居是基于邻居的评分进行预测中至关重要的因素, 如何定义、选取邻居, 将极大地影响推荐系统的最终效果. 一般而言, 利用相似度的大小来定义邻居, 认为两个用户 (或物品) 相似度越大, 他们越相邻. 关于邻居具体的选择标准, 有以下三个基本原则:

- Top-M Filtering: 保留最像 (即相似度最大) 的前 M 个.
- Threshold Filtering: 保留相似度 (绝对值) 大于一个给定阈值 W_{min} 的用户 (或物品).
- Negative Filtering: 去掉不像的用户 (或物品).

(2) 相似度计算

相似度在基于邻居的评分预测算法中既作为寻找邻居的依据发挥作用, 又包含在计算公式当中, 起着非常大的作用.

相似度计算的方式多种多样, 对于可以处理为连续型的评分数据, 可采用 Cosine 相似度和 Pearson 相关系数来度量相似度.

- Cosine 相似度 (Cosine Vector, CV). 对于两个维度相同的列向量 x_a 和 x_b, Cosine 相似度为:

$$cos(x_a, x_b) = \frac{x_a x_b}{\|x_a\| \cdot \|x_b\|}$$

则对于两个用户 u 和 v, 基于评分矩阵 $\boldsymbol{R}$ 定义的 Cosine 相似度为:

$$CV(u,v) = cos(x_u, x_v) = \frac{\sum_{i \in I_{uv}} r_{ui} r_{vi}}{\sqrt{\sum_{i \in I_u} r_{ui}^2 \sum_{j \in I_v} r_{vj}^2}}$$

式中, 分子中的 I_{uv} 表示用户 u 和用户 v 共同打分的物品的集合, 分母中的 I_u 和 I_v 分别表示用户 u 和用户 v 各自打分的物品的集合. 注意, 该公式中分母并不采用 I_{uv} 而是采用 I_u 和 I_v, 这是对原有 Cosine 相似度计算的一种推广.

• Pearson 相关系数 (Pearson Correlation, PC). 对于用户, Pearson 相关系数可以定义为:

$$PC(u,v) = \frac{\sum_{i \in I_{uv}} (r_{ui} - \bar{r}_u)(r_{vi} - \bar{v})}{\sqrt{\sum_{i \in I_{uv}} (r_{ui} - \bar{r}_u)^2 \sum_{i \in I_{uv}} (r_{vi} - \bar{r}_v)^2}}$$

这里的分子与分母都采用 I_{uv}.

类似地, 对于物品, 可以同样定义 Cosine 相似度和 Pearson 相关系数.

对于 0–1 型数据, 如超市购买物品的例子, 1 表示客户购买了某种商品, 说明客户对该商品有兴趣. 0 虽然表示没购买, 却无法区分是不感兴趣, 还是感兴趣只是这次没购买 (如果是这种情形, 恰恰需要推荐). 所以需要定义一种新的相似性, 这里介绍 Jaccard 指数 (Jaccard index)(Hahsler, 2009):

$$sim_{Jaccard}(X,Y) = \frac{|X \cap Y|}{|X \cup Y|}$$

式中, X 为用户 u 取值为 1 的商品的集合 (即购买的商品的集合); Y 为用户 v 取值为 1 的物品的集合; $|X \cap Y|$ 为用户 u 和 v 同时取值为 1 的物品的个数; $|X \cup Y|$ 为用户 u 或者 v 取值为 1 的物品的个数.

除前面的方法以外, 针对不同的问题, 还可以使用其他相似度的测量方法. 比如, 平均平方差异倒数定义为:

$$MSD(u,v) = \frac{|I_{uv}|}{\sum_{i \in I_{uv}} (r_{ui} - r_{vi})^2}$$

Spearman 秩相关系数 (Spearman Rank Correlation, SRC) 定义如下: 令 $\bar{k}_{ui}, \bar{k}_{vi}$ 表示物品 i 在用户 u, v 评价过的物品列表中按得分大小排序后的秩.

$$SRC(u,v) = \frac{\sum_{i \in I_{uv}} (k_{ui} - \bar{k}_u)(k_{vi} - \bar{k}_v)}{\sqrt{\sum_{i \in I_{uv}} (k_{ui} - \bar{k}_u)^2 \sum_{i \in I_{uv}} (k_{vi} - \bar{k}_v)^2}}$$

式中, $\bar{k}_u, \bar{k}_v$ 为用户 u, v 的评分的平均秩 (排序).

上面的相似度度量方法在实际使用过程中有一个普遍的缺陷, 即计算数值的大小不会反映两个人共同打分的物品个数 (或同时为两个物品打分的人的数目) 的影响, 从而在比较过程中缺乏统一的衡量标准. 例如, 用户 a 和用户 b 之间的相似度为 1, 用户 a 和用户 c 之间的相似度为 0.9. 看起来用户 b 比用户 c 更像用户 a 的邻居, 但是实际上, 用户 a 与用户 b 之间只共同打分过 2 个物品, 恰巧打分一致; 而用户 a 与用户 c 共同打分过 200 个物品, 显然用户 c 比用户 b 更可能成为用户 a 的邻居.

因此, 应对相似性权重增加一个基于共同评分物品 (或同时评分的人) 的数目的惩罚:

$$\omega'_{uv} = \frac{\min\{|I_{uv}|, \gamma\}}{\gamma} \times \omega_{uv}$$

$$\omega'_{ij} = \frac{\min\{|U_{ij}|, \gamma\}}{\gamma} \times \omega_{ij}$$

式中, $|I_{uv}|$ 为用户 u 与用户 v 共同评价过的物品数目; $|U_{ij}|$ 表示同时评价过物品 i 与物品 j 的人的数目; γ 是一个事先制定的阈值, 如果 $|I_{uv}|$ (或 $|U_{ij}|$) 小于此值, 则表示相应的权重 ω_{uv} (或 ω_{ij}) 需要调整.

另一种压缩的方法如下:

$$\omega'_{uv} = \frac{|I_{uv}|}{|I_{uv}| + \beta} \times \omega_{uv}$$

$$\omega'_{ij} = \frac{|U_{ij}|}{|U_{ij}| + \beta} \times \omega_{ij}$$

式中, β 为一个给定的压缩参数, 取值大于或等于 0. 若 $\beta = 0$, 则权重完全没有变化. 若 $\beta > 0$, 如果 $|U_{ij}| \gg \beta$, 则表示权重 $\omega'_{uv} \approx \omega_{uv}$, 无须太大调整; 反之, 则权重调整较大.

方差很小的评分实际上信息量不大, 用来调整它的方法叫 Inverse User Frequency, 与 Inverse Document Frequency (IDF) 很像. 对于物品 i:

$$\lambda_i = \log \frac{|U|}{|U_i|}$$

式中, $|U|$ 为用户总数, 即 N; $|U_i|$ 为评价过物品 i 的用户的数目. 用 λ_i 对相关系数进行加权, 得到加权相关系数 (Frequency-Weighted Pearson Correlation, FWPC):

$$FWPC(u, v) = \frac{\sum_{i \in I_{uv}} \lambda_i (r_{ui} - \bar{r}_u)(r_{vi} - \bar{r}_v)}{\sqrt{\sum_{i \in I_{uv}} \lambda_i (r_{ui} - \bar{r}_u)^2 \sum_{i \in I_{uv}} (r_{vi} - \bar{r}_v)^2}}$$

可以用这种方式对物品之间的相似性进行重新定义.

(3) 评分标准化

不同用户的评分范围往往不同, 不同物品的评分范围也往往不同. 因此, 为了更好地预测出每个用户对每个物品的评分, 需要引入标准化函数 h, 下面提供几种标准化函数 h 的选取方法.

- 用户均值中心化 (User-Mean Centered).

$$h(r_{ui}) = r_{ui} - \bar{r}_u$$

$$\hat{r}_{ui} = \bar{r}_u + \frac{\sum\limits_{v \in N_i(u)} \omega_{uv}(r_{vi} - \bar{r}_v)}{\sum\limits_{v \in N_i(u)} |\omega_{uv}|}$$

- 物品均值中心化 (Item-Mean Centered).

$$h(r_{ui}) = r_{ui} - \bar{r}_i$$

$$\hat{r}_{ui} = \bar{r}_i + \frac{\sum\limits_{j \in N_u(i)} \omega_{ij}(r_{uj} - \bar{r}_j)}{\sum\limits_{j \in N_u(i)} |\omega_{ij}|}$$

式中, $\bar{r}_u$ 为用户 u 所有评分的平均值; $\bar{r}_i$ 表示物品 i 获得的所有评分的平均值.

- Z-评分归一化.

用户均值中心化评分:

$$h(r_{ui}) = r_{ui} - \bar{r}_u$$

$$\hat{r}_{ui} = \bar{r}_u + s_u \frac{\sum\limits_{v \in N_i(u)} \omega_{uv}(r_{vi} - \bar{r}_v)/s_v}{\sum\limits_{v \in N_i(u)} |\omega_{uv}|}$$

物品均值中心化评分:

$$h(r_{ui}) = \frac{r_{ui} - \bar{r}_i}{s_i}$$

$$\hat{r}_{ui} = \bar{r}_i + s_i \frac{\sum\limits_{j \in N_u(i)} \omega_{ij}(r_{uj} - \bar{r}_j)/s_j}{\sum\limits_{j \in N_u(i)} |\omega_{ij}|}$$

式中, s_u 和 s_v 为用户 u 和 v 所有评分的标准差; s_i 和 s_j 表示物品 i 和 j 获得的所有评分的标准差, 它们可以用数据进行估计.

> **思考**:
> - 相较于 Cosine 相似度, Pearson 相关系数有什么特点?

2.1.3 网络模型

1. 预备知识

通常用 $G = (V, E)$ 来表示一个网络, 也称为图. 其中, V 为顶点, E 为连接任意两个顶点的边. 图分为有向和无向两种, 即如果每一条边连接的两个顶点有顺序之分, 则称为有向图, 如铁路交通图; 反之为无向图, 如世界地图. 图中的每条边可能是有权重的, 此时将图表示为 $G = (V, E, W)$, 其中, W 为每条边对应的权重; 也可能是没有权重的, 即连接任意两个点 a, b 的边的权重都相等. 关于网络图的几个基本概念如下:

- 点 (Node). 每个网络中的一员称为节点, 可以是个人、组织、网络 ID 等, 用 i 表示, 共 N 个, $i \in \{1, 2, \cdots, N\}$.
- 边 (Edge). 图中的一条线就是一条边, 表示个体间的相互关系. 分为有向边和无向边两种. 用 $a_{ij} = 1$ 表示存在一条从节点 i 到 j 的边, 在无向的情况下, $a_{ij} = a_{ji}$.
- 图 (Graph). 由节点和边组成. 根据边的定义, 分为有向图和无向图.
- 邻接矩阵 (Adjacency Matrix). 用矩阵 $A_{N \times N}$ 来表示, 矩阵中的元素 $a_{ij}(i = 1, 2, \cdots, N; j = 1, 2, \cdots, N)$ 表示从点 i 到点 j 是否有边, $a_{ij} = 0$ 表示没有边, $a_{ij} \neq 0$ 表示有边. 元素可以表示有无边, 也可以表示边的权重. 无向图的邻接矩阵是对称阵, 有向图一般为非对称阵.
- 密度 (Density). 网络中实际存在的边的数目与可能存在的边的数目的比值, 刻画了网络的紧密程度.
- 节点的度 (Degree). 节点的度是指和该节点相关联的边的条数, 又称关联度, 反映网络中点的活跃程度. 对于有向图, 节点的入度是指进入该节点的边的条数; 节点的出度是指从该节点出发的边的条数.
- 路径 (Path). 从节点 i 出发, 经过一条或更多条边, 到达节点 j, 称这些边按顺序相连形成了一条 i 与 j 之间的路径. 包含边数最少或权值加和最小的路径称为最短路径 (shortest path).
- 平均最短路径长度 (Average Shortest Path Length). 对于一个网络而言, 将所有点两两之间的最短路径长度进行算术平均, 得到的就是所谓平均最短路径, 它可以用来衡量网络中点之间的平均距离.
- 网络直径 (Diameter of a Network). 网络图的另一个度量标准, 被定义为网络中最短路径的最大值. 换句话说, 首先计算每个节点到其他节点的最短路径, 网络直径就是最短路径的最大值.

2. Dijkstra 算法介绍

Dijkstra 算法解决的是有向有权图中从某一个指定顶点到其余各顶点的最短路径问题, 前提是图中的权重不允许出现负值. 每一条路径 p 的权重 $\omega(p)$ 就是这条路径上所有边的权重之和, 当从顶点 s 出发, 到达顶点 t 的最短路径存在时, 可以表示为 $\delta(s, t) \min\{\omega(p) : s \to t\}$. 令集合 S 中包含所有已经到达的点, $Q = V - S$ 中包含所有尚未到达的点, 则 Dijkstra 算法从顶点 s 开始, 设顶点 s 对应的距离为 0, 集合中其他所有点对应的距离均为 ∞, 在每一步中更新所到达的点对应的距离. 具体来说, 首先更新顶点 s 的邻近点集合中所有点对应的距离, 找到其中离 s 最近的一点 u_1; 更新 S 和 Q; 以 u_1 为起点, 更新 u_1 的邻近点集合中的距离, 并在其中找到离 u_1 最近的一点 v_1; 比较 $s \to u_1 \to v_1$ 的距离与其他已知的从 s 到 v_1 的距离, 更新 s 到 v_1 的距离为其中最短的一段距离; 多次迭代直到所有的点被遍历为止. Dijkstra 算法是目前已知的最快的单源最短路径算法.

2.2 深度学习

2.2.1 机器翻译模型

在本节, 我们将介绍机器翻译系统的理论算法和技术实践.

2013—2014 年, 之前不温不火的自然语言处理 (NLP) 领域发生了翻天覆地的变化, 因为谷歌大脑的 Mikolov 等人提出了大规模的词嵌入技术 word2vec, RNN、CNN 等深度网络也开始应用于 NLP 的各项任务, 全世界 NLP 研究者欢欣鼓舞、跃跃欲试, 准备告别令人煎熬的平淡期, 开启一个属于 NLP 的新时代.

在这两年机器翻译领域同样发生了大事件. 2013 年牛津大学的 Nal Kalchbrenner 和 Phil Blunsom 提出端到端神经机器翻译编码器 – 解码器模型 (Encoder-Decoder 模型), 2014 年谷歌公司的 Ilya Sutskerver 等人将 LSTM 引入 Encoder-Decoder 模型. 这两件事标志着以神经网络作为基础的机器翻译 (NMT) 开始全面超越此前以统计模型为基础的统计机器翻译 (SMT), 并快速成为在线翻译系统的主流标配. 2016 年谷歌部署神经机器翻译系统 (GNMT) 之后, 互联网上有这样一句广为流传的话: "作为一名翻译, 看到这个新闻的时候, 我理解了 18 世纪纺织工人看到蒸汽机时的忧虑与恐惧."

2015 年, 注意力机制和基于记忆的神经网络突破了 Encoder-Decoder 模型的信息表示瓶颈, 是神经网络机器翻译优于经典的基于短语的机器翻译的关键. 2017 年, 谷歌公司的 Ashish Vaswani 等人参考注意力机制提出了基于自注意力机制的 Transformer 模型, Transformer 家族至今依然在 NLP 的各项任务中保持最佳效果. 近十年 NMT 的发展主要历经了三个阶段: 一般的 Encoder-Decoder 模型、注意力机制模型、Transformer 模型.

下面将逐步深入解析这三个阶段的 NMT, 从而了解业界如何打造工业级 NMT 系统.

1. 新的曙光: Encoder-Decoder 模型

一个自然语言的句子可被视作一个时间序列数据, 类似 LSTM、GRU 等循环神经网络比较适于处理有时间顺序的序列数据. 如果假设把源语言和目标语言都视作一个独立的时间序列数据, 那么机器翻译就是一个序列生成任务. 如何实现一个序列生成任务呢? 一般以循环神经网络为基础的编码器 – 解码器模型框架 (Sequence to Sequence, Seq2Seq) 来做序列生成, 如图 2.1 所示. Seq2Seq 模型包括两个子模型: 一个编码器 (Encoder) 和一个解码器 (Decoder), 编码器、解码器是各自独立的循环神经网络, 该模型首先使用编码器将给定的一个源语言句子映射为一个连续、稠密的向量, 然后再使用一个解码器将该向量转化为一个目标语言句子.

编码器对输入的源语言句子进行编码, 通过非线性变换转化为中间语义表示 C:

$$C = F(X_1, X_2, \cdots, X_m)$$

在第 i 时刻解码器根据句子编码器输出的中间语义表示 C 和之前已经生成的历史信息 $y_1, y_2, \cdots, y_{i-1}$ 来生成下一个目标语言的单词：

$$y_i = G(C, y_1, y_2, \cdots, y_{i-1})$$

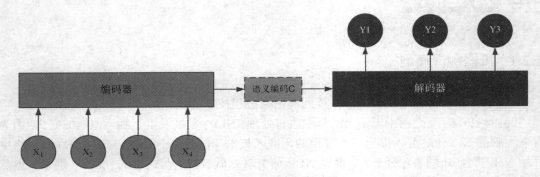

图 2.1 编解码器示意图

每个 y_i 都依次这样产生，即 Seq2Seq 模型就是根据输入源语言句子生成了目标语言句子的翻译模型。源语言与目标语言的句子虽然语言、语序不一样，但具有相同的语义，编码器在将源语言句子浓缩成一个嵌入空间的向量 C 后，解码器能利用隐含在该向量中的语义信息来重新生成具有相同语义的目标语言句子。总而言之，Seq2Seq 神经翻译模型可模拟人类做翻译的两个主要过程：

- 编码器解译来源文字的文意；
- 解码器重新编译该文意至目标语言。

2. 突破飞跃：注意力机制模型

(1) Seq2Seq 模型的局限性

Seq2Seq 模型的一个重要假设是编码器可把输入句子的语义全都压缩成一个固定维度的语义向量，解码器利用该向量的信息就能重新生成具有相同意义但不同语言的句子。由于随着输入句子长度的增加编解码器的性能急剧下降，以一个固定维度中间语义向量作为编码器输出会丢失很多细节信息，因此循环神经网络难以处理输入的长句子，一般的 Seq2Seq 模型存在信息表示的瓶颈。

一般的 Seq2Seq 模型把源语句跟目标语句分开进行处理，不能直接地建模源语句跟目标语句之间的关系。那么如何解决这种局限性呢？2015 年，Bahdanau 等人发表论文首次把注意力机制应用到联合翻译和对齐单词中，解决了 Seq2Seq 模型的瓶颈问题。注意力机制可计算目标词与每个源语词之间的关系，从而直接建模源语句与目标语句之间的关系。注意力机制又是什么神器，可让 NMT 一战成名决胜机器翻译竞赛呢？

(2) 注意力机制的一般原理

通俗地解释，在数据库里一般用主键 (Key) 唯一地标识某一条数据记录 (Value)，访问某一条数据记录的时候可用查询语句 Query 搜索与查询条件匹配的主键并取出其中的数据。注意力机制类似于该思路，是一种软寻址的概念：假设数据按照 $<Key, Value>$ 存储，计算所有的主键与某一个查询条件的匹配程度，作为权重值再分别与各条数据做加权和作为查询的结果，该结果即注意力。

图 2.2 描述了注意力机制的一般原理。

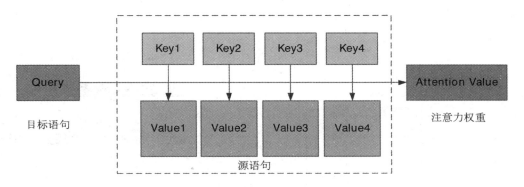

图 2.2 注意力机制的一般原理

首先，将源语句中的构成元素想象成由一系列的 < Key, Value > 数据对构成，目标语句由一系列元素 (Query) 构成；然后给定目标语句中的某个元素，通过计算元素和各个主键的相似性或者相关性，得到每个主键对应数据记录的权重系数；最后，可对数据记录进行加权，得到最终的注意力权重。因此，本质上注意力机制是对源语句中元素的数据记录值进行加权求和，而元素和主键用来计算对应数据记录的权重系数。一般性计算公式为：

$$Attention(Query, Source) = \sum_{i=1}^{L_x} Similarity(Query, Key_i) \cdot Value_i$$

在机器翻译中，Seq2Seq 模型一般由多个 LSTM/GRU 等 RNN 层叠起来。2016 年 9 月谷歌发布神经机器翻译系统 GNMT，采用 Seq2Seq+ 注意力机制的模型框架，编码器网络和解码器网络都具有 8 个 LSTM 隐层，编码器的输出通过注意力机制加权平均后输入解码器的各个 LSTM 隐层，最后连接 Softmax 层输出每个目标语言词典的每个词的概率 (见图 2.3)。
GNMT 如何计算让性能大幅提升的注意力呢？假设 (X,Y) 为平行语料的任一组源语句 – 目标语句对，则：

- 源语句长度为 M 的字符串：$X = x_1, x_2, \cdots, x_M$
- 目标语句长度为 N 的字符串：$Y = y_1, y_2, \cdots, y_N$
- 编码器输出 d 维向量 h 作为 X 的编码：$h_1, h_2, \cdots, h_M = F(x_1, x_2, \cdots, x_M)$

利用贝叶斯定理，句子对的条件概率：$P(Y|X) = \prod_{i=1}^{N} P(y_i|y_0, y_1, \cdots, y_{i-1}; h_1, h_2, \cdots, h_M)$，

解码时解码器在时间点 i 根据编码器输出的编码和前 $i-1$ 个解码器输出，最大化 $P(Y|X)$ 可求得目标词。

GNMT 注意力机制实际的计算步骤如下：

$$s_t = Attention(y_{i-1}, x_t) \quad \forall t, 1 \leqslant t \leqslant M$$

$$p_t = \frac{\exp(s_t)}{\sum_{t-1}^{M} \exp(s_t)} \quad \forall t, 1 \leqslant t \leqslant M$$

$$a_i = \sum_{t=1}^{M} p_t \cdot x_t$$

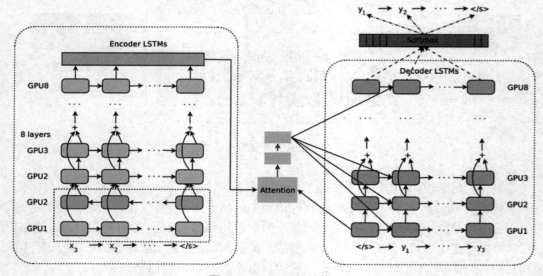

图 2.3 GNMT 结构

3. 高光时刻: 基于自注意力机制的 Transformer 模型

上文提到基于 Seq2Seq+ 注意力机制比一般的 Seq2Seq 的模型架构取得了更好的效果, 那么这种组合有什么缺点呢? 事实上循环神经网络存在着一个困扰研究者已久的问题: 无法有效地平行运算, 但不久研究者就等来了突破. 2017 年 6 月 Transformer 模型被提出, 当时谷歌在发表的一篇论文 (Vaswani A, et al. 2017) 里参考了注意力机制, 提出了自注意力机制 (Self-Attention) 及新的神经网络结构 —— Transformer. 该模型具有以下优点:

• 传统的 Seq2Seq 模型以 RNN 为主, 制约了 GPU 的训练速度, Transformer 模型是一个完全不用 RNN 和 CNN 的可并行机制计算注意力的模型;

• Transformer 模型改进了 RNN 最被人诟病的训练慢的缺点, 利用自注意力机制实现快速并行计算, 并且 Transformer 模型可以增加到非常深的深度, 充分发掘 DNN 模型的特性, 提升模型准确率.

下面我们深入解析 Transformer 模型架构.

(1) Transformer 模型架构

Transformer 模型本质上也是一个 Seq2Seq 模型, 由编码器、解码器和它们之间的连接层组成, 如图 2.4 所示. 在上文提到的谷歌论文中介绍的 The Transformer 编码器由 $N = 6$ 个完全相同的编码层 (Encoder Layer) 堆叠而成, 每一层都有两个子层. 第一个子层是一个 Multi-Head Attention 机制, 第二个子层是一个简单的、位置完全连接的前馈网络 (Feed-Forward

Network). 我们对每个子层再采用一个残差连接 (Residual Connection),接着进行层归一化 (Layer Normalization). 每个子层的输出是 $LayerNorm(x + Sublayer(x))$,其中 $Sublayer(x)$ 是由子层本身实现的函数.

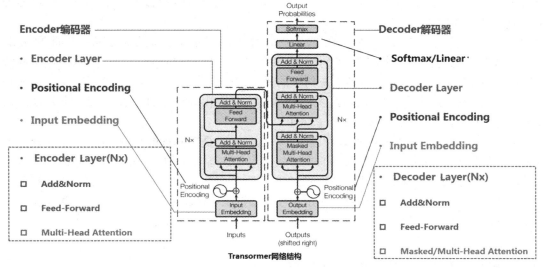

图 2.4 Transformer 网络结构

解码器同样由 $N = 6$ 个完全相同的解码层 (Decoder Layer) 堆叠而成. 除了与每个编码器层中相同的两个子层之外,解码器还插入第三个子层 (Encoder-Decoder Attention 层),该层对编码器堆栈的输出执行 Multi-Head Attention. 与编码器类似,我们在每个子层再采用残差连接,然后进行层归一化.

Transformer 模型计算注意力的方式有三种:
- 编码器自注意力,每一个编码器都有 Multi-Head Attention 层;
- 解码器自注意力,每一个编码器都有 Masked/Multi-Head Attention 层;
- 编码器 – 解码器注意力,每一个解码器都有一个 Encoder-Decoder Attention, 过程和过去的 Seq2Seq+Attention 的模型相似.

(2) 自注意力机制

Transformer 模型的核心思想就是自注意力机制,能注意输入序列的不同位置以计算该序列的表达能力. 自注意力机制, 顾名思义指的不是源语句和目标语句之间的注意力机制, 而是同一个语句内部元素之间发生的注意力机制. 在计算一般 Seq2Seq 模型中的注意力时以解码器的输出作为查询向量 q, 以编码器的输出序列作为键向量 k、值向量 v, Attention 机制发生在目标语句的元素和源语句中的所有元素之间.

自注意力机制的计算过程是将编码器或解码器的输入序列的每个位置的向量通过 3 个线性转换分别变成 3 个向量: 查询向量 q、键向量 k、值向量 v,并将每个位置的 q 拿去跟序列中其他位置的 k 做匹配, 算出匹配程度后利用 softmax 层取得介于 0 到 1 之间的权重值, 并以此权重跟每个位置的 v 作加权平均, 最后取得该位置的输出向量 z. 下面介绍自注意力的计算方法.

1) 可缩放的点积注意力

可缩放的点积注意力即如何使用向量来计算自注意力, 通过四个步骤来计算自注意力:

第一步, 从每个编码器的输入向量 (每个单词的词向量) 生成三个向量: 查询向量 q、键向量 k、值向量 v. 矩阵运算中这三个向量是通过编解码器输入 X 与三个权重矩阵 W^q、W^k、W^v 相乘创建的.

第二步, 计算得分. 如图 2.5 所示, 输入一个句子 "Thinking Machine", 第一个词 "Thinking" 计算自注意力向量, 需将输入句子中的每个单词对 "Thinking" 打分. 分数决定了在编码单词 "Thinking" 的过程中有多重视句子的其他部分. 分数是通过打分单词 (所有输入句子的单词) 的键向量 k 与 "Thinking" 的查询向量 q 相点积来计算的. 比如, 第一个分数是 q_1 和 k_1 的点积, 第二个分数是 q_1 和 k_2 的点积.

第三步, 缩放求和. 将分数乘以缩放因子 $1/\sqrt{d_k}$ (d_k 是键向量的维数 $d_k = 64$) 让梯度更稳定, 然后通过 softmax 传递结果. softmax 的作用是使所有单词的分数归一化, 得到的分数都是正值且和为 1. softmax 分数决定了每个单词对编码当下位置 ("Thinking") 的贡献.

第四步, 将每个值向量 v 乘以 softmax 分数, 希望关注语义上相关的单词, 并弱化不相关的单词. 对加权值向量求和, 即得到自注意力层在该位置的输出 z_i.

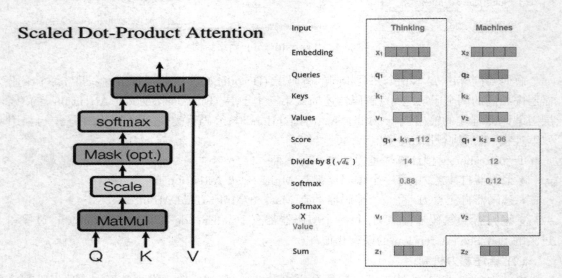

图 2.5 可缩放的点积注意力

因此, 可缩放的点积注意力可通过下面的公式计算:

$$Attention(Q, K, V) = softmax\left(\frac{QK^T}{\sqrt{d_k}}\right)V$$

在实际中, 注意力计算是以矩阵形式完成的, 以便算得更快. 接下来看看如何用矩阵运算实现自注意力机制.

首先求取查询向量矩阵 Q、键向量矩阵 K 和值向量矩阵 V, 通过权重矩阵 W^q、W^k、W^v 与输入矩阵 X 相乘得到; 同样求取任意一个单词的得分是通过它的键向量 k 与所有单词的查询向量 q 的点积来计算的, 可以把所有单词的键向量 k 的转置组成一个键向量矩阵 K^T, 把所有单词的查询向量 q 组合在一起成为查询向量矩阵 Q, 这两个矩阵相乘得到注意力得

分矩阵 $A = QK^T$; 然后, 对注意力得分矩阵 A 求 softmax 得到归一化的得分矩阵 $\hat{A}$, 这个矩阵在左乘以值向量矩阵 V 得到输出矩阵 Z (见图 2.6).

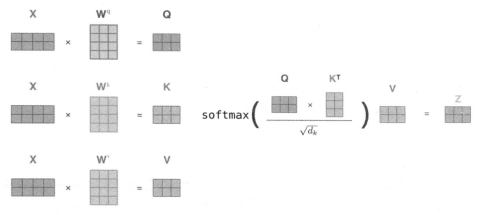

图 2.6 注意力计算矩阵表达

2) 多头注意力

如果只计算一个 Attention, 很难捕捉输入句中所有空间的信息, 为了优化模型, 谷歌论文中提出了一个新颖的做法 —— Multi-Head Attention. Multi-Head Attention 是不能只用嵌入向量维度 d_{model} 的 K, Q, V 做单一 Attention, 而是把 K, Q, V 线性投射到不同空间 h 次, 分别变成维度 d_q, d_k 和 d_v 再各自做 Attention.

其中, $d_q = d_k = d_v = d_{model}/h = 64$ 就是投射到 h 个 Head 上. Multi-Head Attention 允许模型的不同表示子空间联合关注不同位置的信息, 如果只有一个 Attention Head 则它的平均值会削弱这个信息.

Multi-Head Attention 为每个 Head 保持独立的查询/键/值权重矩阵 W_i^Q, W_i^K, W_i^V, 从而产生不同的查询/键/值矩阵 Q_i, K_i, V_i. 用 X 乘以 W_i^Q, W_i^K, W_i^V 矩阵来产生查询/键/值矩阵 Q_i, K_i, V_i. 与上述的自注意力计算相同, 只需 8 次不同的权重矩阵运算可得到 8 个不同的 Z_i 矩阵, 每一组都代表将输入文字的隐向量投射到不同空间. 最后把这 8 个矩阵拼在一起, 通过乘上一个权重矩阵 W^O, 还原成一个输出矩阵 Z.

Multi-Head Attention 的每个 Head 到底关注句子中的什么信息呢? 不同的注意力的 Head 集中在哪里? 以下面这两句话为例 "The animal didn't cross the street because it was too tired" 和 "The animal didn't cross the street because it was too wide", 两个句子中 "it" 指的是什么呢? 是 "street" 还是 "animal"? 当我们编码 "it" 一词时, it 的注意力集中在 "animal" 上和 "street" 上, 从某种意义上说, 模型对 "it" 一词的表达在某种程度上是 "animal" 和 "street" 的代表, 第一句的 it 更强烈地指向 animal, 第二句的 it 更强烈地指向 street.

(3) Transformer 模型其他结构
1) 残差连接与归一化

编解码器有一种特别的结构:Multi-Head Attention 的输出接到 Feed-Forward Layer 之间有一个子层: Residual Connection 和 Layer Normalization(LN), 即残差连接与层归一化. 残差连接是构建一种新的残差结构, 将输出改写为输入的残差, 使得模型在训练时, 微小的变化可以被注意到, 该方法在计算机视觉领域常用.

在把数据送入激活函数之前需进行归一化, 因为我们不希望输入数据落在激活函数的饱和区. LN 是深度学习中的一种正规化方法, 一般和 Batch Normalization(BN) 进行比较. BN 的主要思想就是在每一层的每一批数据上进行归一化, LN 是在每一个样本上计算均值和方差, LN 的优点在于独立计算并针对单一样本进行正规化, BN 则是在批方向计算均值和方差.

2) 前馈神经网络

编解码层中的注意力子层输出都会接到一个全连接网络: 前馈网络 (Feed-Forward Networks, FFN), 包含两个线性转换和一个 ReLu, 谷歌论文是根据各个位置 (输入句中的每个文字) 分别做 FFN, 因此称为 Point-Wise 的 FFN. 计算公式如下:

$$FFN = \max(0, xW_1 + b_1)W_2 + b_2$$

3) 线性变换和 softmax 层

解码器最后会输出一个实数向量. 把浮点数变成一个单词, 便是线性变换层要做的工作, 它之后就是 softmax 层. 线性变换层是一个简单的全连接神经网络, 它可以把解码器产生的向量投射到一个比它大得多的、被称作对数概率 (logits) 的向量里.

不妨假设我们的模型从训练集中学习 10 000 个不同的英语单词 (模型的 "输出词表"). 因此 logit 向量为 10 000 个单元格长度的向量——每个单元格对应某一个单词的分数. 接下来的 softmax 层便会把那些分数变成概率 (都为正数, 上限 1.0). 概率最高的单元格被选中, 并且它对应的单词被作为这个时间步的输出.

4) 位置编码

Seq2Seq 模型的输入仅仅是词向量, Transformer 模型摒弃了循环和卷积, 无法提取序列顺序的信息, 如果缺失了序列顺序信息, 可能会导致所有词语都对了, 但是无法组成有意义的语句. 谷歌论文的作者是怎么解决这个问题呢? 为了让模型利用序列的顺序, 必须注入序列中关于词语相对或者绝对位置的信息. 在论文中作者引入位置编码 (Positional Encoding): 对序列中的词语出现的位置进行编码. 图 2.7 是 20 个词 512 个嵌入维度上的位置编码可视化.

图 2.7　20 个词 512 个嵌入维度上的位置编码可视化

将句子中每个词的位置编码添加到编码器和解码器堆栈底部的输入嵌入中, 位置编码和

词嵌入的维度 d_{model} 相同,所以它们可以相加. 论文使用不同频率的正弦和余弦函数获取位置信息:

$$PE_{(pos,2i)} = \sin(pos/10000^{2i/d_{model}})$$
$$PE_{(pos,2i+1)} = \cos(pos/10000^{2i/d_{model}})$$

式中, pos 为位置; i 为维度, 在偶数位置使用正弦编码, 在奇数位置使用余弦编码. 位置编码的每个维度对应一个正弦曲线.

2.2.2 图像分析模型

图像数据的来源主要是位图, 也称点阵图, 通常以 JPG、BMP、PNG 等格式存储. 位图中彩色的点由 RGB 矩阵表示, RGB 代表光学三原色红、绿、蓝. 通常每种原色可以分成 256 阶, 用 0 ~ 255 的整数表示 (有些软件或者程序包中是除以 255 后的实数). 每张彩色图由 3 个矩阵 (或一个 3 维张量) 构成, 分别代表 R、G、B 的数值.

使用某些加权方式[①] 可以把三通道的彩色图像转化成灰度图, 灰度图只需 1 个矩阵即可描述, 0 代表黑, 255 代表白, 中间的数值代表灰度. 由于其结构简单, 计算方便, 在不需要考虑颜色特征的场景下比较常用.

传统的图像分析方式主要基于算法来提取各种特征, 例如, 先用图像识别的算法提取边缘, 然后提取形状、颜色、纹理等特征, 用数值进行衡量. 每幅图像都能转化成一系列的特征, 很多幅图像在一起就能得到一个矩阵, 从而可以用各种统计和机器学习方法进行分析. 该方式的解释性很好, 容易和其他特征整合在一起, 但太依赖人的经验, 而且并不是所有人类能识别的特征都能显式地量化.

另一种方式是将图像的每一个像素值作为一个特征, 比如某图像分辨率是 100 × 100, 那么可以对应 1 万个特征. 这种对图像进行像素级理解的方式符合人类神经的反应机制, 也很适合使用基于神经网络的深度学习模型. 直接使用深度学习进行图像分类等操作的话, 可以不依赖于人类的经验, 通常准确率也更高, 但解释性不太好.

在实际的应用中, 通常会综合使用深度学习的方法, 再结合一些传统图像特征进行分析, 兼顾预测性和解释性. 最常用的图像特征是纹理特征, 纹理是指影像中大量的规律性很强或者很弱的相似结构或图形结构, 可以理解成影像灰度在空间上的变化与重复. 常用的 Haralick 定义的灰度共生矩阵 (GLCM) 主要包括以下特征:

- 同质度, 也称逆差距 (IDM), 图像的局部灰度越均匀该值越大.
- 对比度 (Contrast), 反映局部灰度变化总量, 差别越大该值越大.
- 熵 (Entropy), 图像信息量的度量, 图像纹理越复杂熵越大.
- 角二阶矩 (ASM), 也称能量, 是图形灰度均匀性的度量, 灰度越均匀该值越大.
- 相关性 (Correlation), 表示灰度值沿某方向的延伸长度, 越长该值越大.
- 方差 (VAR), 描述偏差的程度.
- 和平均值 (Sum Average), 简称 SAV.
- 和方差 (Sum Variance), 简称 SVA.

---
① 例如平均法、基于亮度的加权方法等.

- 差分方差 (Difference Variance), 简称 DVA.
- 和熵 (Sum Entropy), 简称 SEN.
- 差分熵 (Difference Entropy), 简称 DEN.

在图像分析中, 根据任务来划分的话, 主要有图像分类、目标检测、语义分割等, 不同的任务对应不同的深度学习模型, 下面分别进行介绍.

1. 图像分类

图像分类是最常见的图像应用, 通常基于卷积神经网络 (CNN) 模型将图像对应到一维结构的分类标签, 从而实现分类. 通过分析卷积核的权重, 并进行可视化展现, 可以实现一定程度的可解释性. 对于预测结果, 可以使用准确率、灵敏度、特异性、精度、召回、F1 得分、AUC 等常见指标进行评价.

(1) LeNet

图灵奖得主 LeCun 在其官网提供了 MINIST 手写数字数据集和 LeNet-5 模型, 这是早期比较经典的图像分类模型, 常用于教学和简单的图像分类. LeNet-5 共有 7 层, 如图 2.8 所示, 在 MINIST 数据集上, 图像分辨率为 32×32. 第 1 层是卷积层, 输入图像的维度为 32×32. 设置 6 个 5×5 的卷积核, 采用默认的步幅 1, 卷积结束后, 使用 tanh 激活函数进行激活.

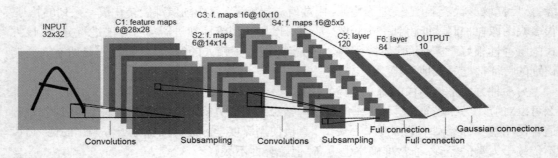

图 2.8 LeNet-5 结构图

第 2 层为池化层, 设置采样区域为 2×2, 每个维度的步幅都为 2, 进行最大池化, 得到的结果是 14×14 的特征图. 第 3 层是卷积层, 使用 16 个 5×5 的卷积核进行卷积, 默认步幅在两个维度都为 1, 卷积后使用 tanh 激活函数进行激活. 第 4 层也是池化层, 和第 2 层采用相同的参数.

第 5 层是卷积全连接层, 使用 120 个 5×5 的卷积核对上一层的 16 个特征图进行全连接的卷积操作. 第 6 层是全连接层, 具有 84 个神经元, 和第 5 层进行全连接, 仍然使用 tahn 函数激活. 第 7 层是全连接输出层, 关联到类比标签.

LeNet-5 结构简单 (Lecun Y, Bottou L, Bengio Y, et al., 1998), 层数也不多, 运算性能很好, 在手写数字识别上得到了很好的结果, 在一些数据量不太大的场景下应用比较广泛.

(2) AlexNet

AlexNet 模型的名字来源于其第一设计者 Alex krizhevsky 的姓名,该模型 2012 年赢得了 ImageNet 图像识别挑战赛冠军. 其结构如图 2.9 所示, 包括 5 层卷积和 2 层全连接隐层, 以及 1 层全连接输出层.

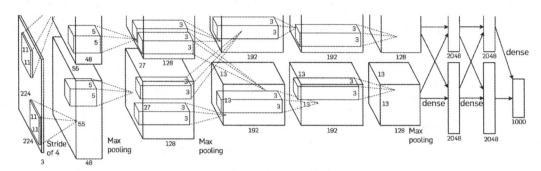

图 2.9 AlexNet 结构图

第 1 层卷积核的维度为 11×11, 步幅为 4, 适用于分辨率相对较高的图像. 第 2 层是采样区域为 3×3、步幅为 2 的最大池化层. 第 3 层是 5×5 的卷积层, 接着一个采样区域为 3×3、步幅为 2 的最大池化层. 此后连续 3 个 3×3 的卷积层, 再接一个 3×3 的最大池化层.

在 5 层卷积之后是 2 个全连接层, 神经元个数都为 4 096, 最后是一个全连接输出层. 需要注意的是, 初始论文中使用了双数据流的设计, 如图 2.9 所示, 分成了上下两部分 (每一部分全连接层的神经元个数为 2 048), 现在随着显卡性能的提高, AlexNet 直接实现以上的结构, 无需早期的这种设计.

AlexNet 模型的参数数目非常巨大, 接近 1GB. 其激活函数使用 ReLU, 训练起来更容易, 此外还使用丢弃法 (Dropout) 来控制全连接的复杂度. 相比 LeNet, AlexNet 可以处理更大的图像和更多的数据, 一直很受欢迎.

(3) VGG

VGG 模型的名字来源于论文作者所在的实验室 Visual Geometry Group, 提出了可以通过重复使用简单的基础块来构建深度模型的思路 (Simonyan K, Zisserman A, 2014). 一个基本思路是连续使用个数相同的填充为 1、窗口形状为 3×3 的卷积层后接上一个步幅为 2、窗口形状为 2×2 的最大池化层. 不同组合方式对应不同网络. 图 2.10 显示了几种常用的结构.

我们以最左侧的 "A" 模型为例, 也称 VGG-11, 包含了 5 个卷积块, 图 2.10 中 "conv3-64" 表示一个卷积层, 包含 64 个 3×3 的卷积核. 可以看出, 前 2 个卷积块使用了单卷积层, 卷积核的数目分别是 64 和 128, 后 3 个卷积块使用了双卷积层, 卷积核的数目分别是 256、512 和 512.

此后接了 3 个全连接层, 神经元的数目分别为 4 096、4 096 和 1 000, 由于 5 个卷积块一共包含 8 个卷积层, 该模型的卷积层数目为 11, 因此称为 VGG-11. 用类似的方式可以构造 VGG-16、VGG-19, 都是常用的模型. VGG 的这种方式可以基于重复的基础块来构造模型, 为用户提供了很高的灵活度.

(4) ResNet

残差网络 ResNet 在 2015 年的 ImageNet 图像识别挑战赛中夺冠, 它通过残差块以短路

ConvNet Configuration					
A	A-LRN	B	C	D	E
11 weight layers	11 weight layers	13 weight layers	16 weight layers	16 weight layers	19 weight layers
input (224 × 224 RGB image)					
conv3-64	conv3-64 LRN	conv3-64 conv3-64	conv3-64 conv3-64	conv3-64 conv3-64	conv3-64 conv3-64
maxpool					
conv3-128	conv3-128	conv3-128 conv3-128	conv3-128 conv3-128	conv3-128 conv3-128	conv3-128 conv3-128
maxpool					
conv3-256 conv3-256	conv3-256 conv3-256	conv3-256 conv3-256	conv3-256 conv3-256 conv1-256	conv3-256 conv3-256 conv3-256	conv3-256 conv3-256 conv3-256 conv3-256
maxpool					
conv3-512 conv3-512	conv3-512 conv3-512	conv3-512 conv3-512	conv3-512 conv3-512 conv1-512	conv3-512 conv3-512 conv3-512	conv3-512 conv3-512 conv3-512 conv3-512
maxpool					
conv3-512 conv3-512	conv3-512 conv3-512	conv3-512 conv3-512	conv3-512 conv3-512 conv1-512	conv3-512 conv3-512 conv3-512	conv3-512 conv3-512 conv3-512 conv3-512
maxpool					
FC-4096					
FC-4096					
FC-1000					
softmax					

图 2.10 VGG 结构图

连接的方式来解决层数增加后误差增大的问题 (He K, Zhang X, Ren S, et al., 2016). 理论上来看, 在神经网络模型中增加新的层数可以降低训练误差, 因为原模型的解空间只是新模型解的子空间, 新模型在拟合时可能会得到更优的解. 但是在实际的操作中, 当 CNN 的网络层数超过一定的深度后并不能提升准确率, 而且会大大降低计算性能. 例如 VGG-19 再继续增加层数的话, 就会导致分类效果开始变差, 主要原因在于梯度爆炸和梯度消失.

一个解决办法就是 ResNet 残差网络, 如图 2.11 所示. 假设我们希望学习的映射为 $f(x) = \mathcal{F}(x) + x$, 其残差映射即为 $f(x) - x = \mathcal{F}(x)$, 如果增加新层后能将模型训练成恒等映射 $f(x) = x$, 那么新模型将和原模型同样有效, 并且还有可能得到更优的解, 因此增加新层后的模型总体上更优.

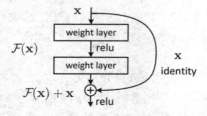

图 2.11 ResNet 结构图

基于这样的思路, 把图 2.11 中加权运算的权重学习成 0 即可. 在实际的应用中, 可以使用图 2.12 所示的两种结构.

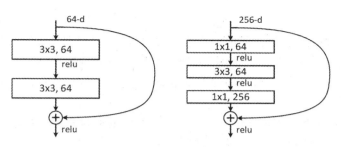

图 2.12 ResNet 的两种残差块

以左侧的残差块结构为例, 它包含了 2 个相同的卷积层, 每个都有 64 个 3×3 的卷积核, 每个卷积层后接 ReLU 激活函数, 此外将输入跳过这 2 个卷积运算后直接加在最后的 ReLU 之前. 实际操作中显示, ResNet 的深度增加到数百层后都能有很好的表现, 这也深刻影响了后来的深度神经网络的设计.

2. 目标检测

图像中经常会包含一个或多个我们感兴趣的目标, 我们不仅想知道它们的类别, 还想得到它们在图像中的具体位置. 在计算机视觉里, 这类任务称为目标检测 (Object Detection) 或物体检测. 在目标检测里, 通常使用真实边界框 (Bounding Box) 来描述目标位置, 用左上角和右下角的坐标表示. 训练时通常在输入图像中采样大量区域, 其边界框称为锚框 (Anchor Box), 然后判断其中是否包含感兴趣的目标, 并调整边缘更精准地预测真实边界框, 从而实现目标检测.

(1) SSD

单发多框检测 (Single Shot Multibox Detection, SSD) 由一个基础网络块和若干个多尺度特征块串联而成 (Liu W, Anguelov D, Erhan D, et al., 2016). 基础块常选用经典的 CNN 模型, 例如 VGG、ResNet 等, 然后基于基础块和多尺度特征块的特征图生成锚框, 如图 2.13 所示.

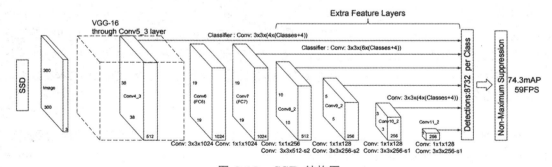

图 2.13 SSD 结构图

在 SSD 中, 采用 VGG-16 作为基础模型, 然后新增具有不同特征图的卷积层来进行检测, 由于采用了大小不同的特征图, 因此可以实现多尺度的检测. 通常前面的特征图尺度较大, 然后通过步幅和池化操作来减小特征图, 较大的特征图可以用来检测较小的目标, 较小的特征图可以用来检测较大的目标. 此外, SSD 还会为每个预测单元的边界框设置先验框, 从而预测其尺度和长宽比. 目标检测的过程中包含两个损失, 分别是锚框类别损失和锚框偏移量的损失, 在 SSD 中采用交叉熵和 L_1 范数的和来定义损失函数.

SSD 的准确性不错, 对小目标的预测效果也很好, 在图像目标检测中有着广泛应用.

(2) R-CNN

区域卷积神经网络 (Region-Based CNN, R-CNN) 是将深度模型应用于目标检测的开创性工作之一 (Girshick R, Donahue J, Darrell T, et al., 2014), 其改进版包括快速的 R-CNN (Fast R-CNN)、更快的 R-CNN (Faster R-CNN)、掩码 R-CNN (Mask R-CNN) 等. 图 2.14 显示了该模型的结构.

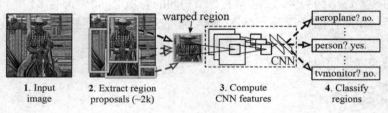

图 2.14　R-CNN 结构图

先使用选择性搜索得到提议区域 (Region Proposals), 每个提议区域都会被标注类别和边界框. 再选取一个预训练的 CNN, 在输出层之前截断, 抽取提议区特征. 然后将提议区域的特征及其标注的类别作为一个样本, 训练 SVM 模型对目标分类, 每个 SVM 预测一个类别. 最后将每个提议区域的特征和边界框作为一个样本, 训练线性回归模型来预测真实的边界框.

R-CNN 可以基于 CNN 来抽取图像特征, 在多个候选的提议区域中通过训练搜索到最优的结果, 从而实现目标检测.

3. 语义分割

语义分割 (Semantic Segmentation) 是指将图像分割成属于不同语义类别的区域, 其标注和预测都是像素级. 语义分割只判断类别, 无法判断个体. 还有个相似的概念实例分割 (Instance Segmentation), 它研究如何识别图像中各个目标实例的像素级区域. 此外, 语义分割和传统的图像分割 (Image Segmentation) 也存在区别, 图像分割通常是指将图像分割成若干组成区域, 在训练时不需要有关图像像素的标签信息, 在预测时也无法保证分割出的区域具有我们希望得到的语义.

(1) FCN

全卷积网络 (Fully Convolutional Network, FCN) 采用 CNN 实现了从图像像素到像素类别的变换 (Long J, Shelhamer E, Darrell T, 2015). 其核心思想是通过上采样 (Upsample) 层将中间层特征图的高和宽变换回输入图像的尺寸, 从而令预测结果与输入图像在空间维上一一对应, 如图 2.15 所示.

基于 FCN 模型, 给定图像空间维的位置后, 其通道维的输出结果即为该位置对应像素的类别预测, 从而实现语义分割.

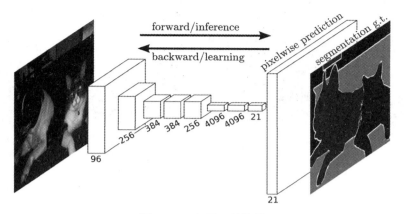

图 2.15　FCN 结构图

(2) U-Net

U-Net 最早用于细胞分割, 在生物医学领域有着广泛的应用 (Ronneberger O, Fischer P, Brox T, 2015). 其基本思路和经典 FCN 类似, 也是一种全卷积网络. 主要特点在于前半部分通过最大池化下采样, 后半部分通过转置卷积层上采样, 形成对称的 U 形结构, 如图 2.16 所示.

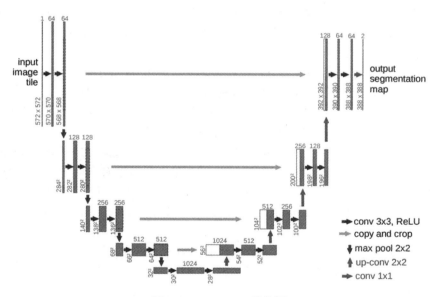

图 2.16　U-Net 结构图

第3章
音乐风格识别

3.1 背景介绍

音乐是现代社会中与我们息息相关的一种数据,我们在看电视、看电影、就餐、坐车等各种场合都能够听到各种音乐,而且我们使用的许多移动智能设备可以存储或播放各式各样的音乐. 伴随着数字音乐的日渐流行以及数字音乐内容的与日俱增,音乐信息检索 (Music Information Retrieval, MIR) 领域的研究引起了更多人的兴趣. MIR 主要包含音乐特征的提取和变换、音乐分类、音乐推荐、音乐搜索等,其中音乐特征的提取是极其基础也是极其重要的工作.

音乐分类 (Music Classification) 是 MIR 领域一个很重要的方向,它主要研究如何为每首音乐分配一个合适的标签 (Label),诸如风格 (Genre)、情感 (Mood)、歌手 (Artist) 甚至乐器 (Instrument). 音乐分类也构成了 MIR 的一个十分重要的功能组件,它不仅有利于数字音乐的高效管理,也为用户端的推荐和搜索带去了极大的便利.

音乐风格的识别是音乐分类中研究得较多的一个方向,其目标就是根据从音乐中提取出的各种音频特征来自动识别该音乐片段的风格. 常见的音乐风格有古典、流行、摇滚等,因而这是一个多分类问题. 如果能构建一个预测音乐风格的高准确率模型,就可以自动化地为库存音乐以及每年更新到曲库的大量音乐打上标签,再使用这些标签分析用户听音乐的喜好,进而改善推荐引擎的推荐效果.

音乐风格识别任务一般来说可以分为图 3.1 中的几个步骤。

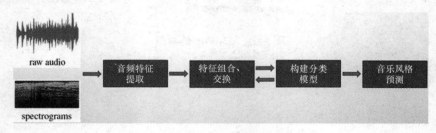

图 3.1 音乐风格识别的主要步骤

首先需要获得原始音频信号,并对其进行频谱分析得到声谱图 (用于描述声音的各频率成分如何随时间变化的热图);再基于音频信号及其声谱图提取重要的音频特征 (如过零率、能量以及梅尔频率倒谱系数等); 初次提取的音频信号可能不足以完成分类任务, 因此有必要对其进行适当的变换或组合, 形成更复杂、更高级同时也更有利于模型进行学习的特征; 在完成了特征的提取之后, 我们就可以在各种音频特征之上, 利用带有音乐风格标签的足量的数据, 训练机器学习模型, 通过调节模型参数、调整模型架构, 甚至更换不同的模型来学习数据背后的统计规律; 模型训练完毕之后, 就可以对库存音乐、未来新入库的音乐以及平台用户和音乐人大量的翻唱和自创歌曲进行预测.

改善音乐分类任务可以从多个角度出发. 首先是挖掘出更多更好的音频特征, 除了多个音频特征以外, 还可以借助一些特征组合变换的手段, 对现有特征进行转化; 接着是选取学习能力较强的模型, 一般来说, 如果音频特征与标签之间的关系并不是简单的线性关系, 那么复杂度较低的线性分类器就无法很好地完成任务, 所以模型层需要挑选诸如树模型或者深度网络这些复杂度较高的可以较好学习非线性关系的模型; 此外, 单个模型的学习能力终究有限, 集成方法 (Boosting) 是综合多个学习器进行预测的典型方法, 也能很好地改善预测性能.

本案例采用了损失敏感 (Cost-Sensitive) 型的 Boosting 算法与逻辑回归 (Logistic Regression) 嵌套的混合动力模型架构来对音乐风格进行预测, 既达到了比单独的 Boosting 算法或逻辑回归更好的预测效果, 还通过调节误判损失矩阵改善了部分类别预测的准确率和召回率, 此外, 混合动力模型架构中的 Boosting 模型天然地完成了对原始音频特征的加工, 每个基分类器 (决策树) 的结构都是强大的非线性转换器, 这给我们提供了十分有效的音频特征编码手段.

本章 3.2 节主要介绍音频数据和音频特征, 以及基于损失敏感型多分类 Boosting 算法和逻辑回归的混合动力模型架构; 3.3 节主要是对本案例采用的音乐数据进行简单的描述分析, 包括音乐数据及其标签的来源、对音乐数据的加工方式、采用的音频特征和对音频特征的单特征分析; 3.4 节主要展示并对比了逻辑回归、Boosting 算法和混合动力模型架构在实际音乐风格分类任务中的预测效果, 以及在工程上进行优化和加速的手段.

3.2 方法简介

3.2.1 音频数据和音频特征

一首音乐其实就是一个音频文件. 音频数据的存储形式和展现形式大都是一串时序数据, 记录着振幅随时间的变化. 图 3.2 是一个十分典型的音频声波图, 横轴是时间, 纵轴是振幅 (中心化后).

音乐不仅数量繁多, 风格也很多样, 如古典音乐、电子音乐、蓝调、爵士音乐、摇滚音乐、流行音乐和世界音乐等. 每种风格的音乐都有其特点和历史渊源.

- 古典音乐 (Classical Music), 这里指从西方中世纪开始至今的、在欧洲主流文化背景下创作的音乐, 主要因其复杂多样的创作技术和所能承载的厚重内涵而有别于通俗音乐和民间音乐, 同时也是 1750—1820 年这一段时间的欧洲主流音乐.

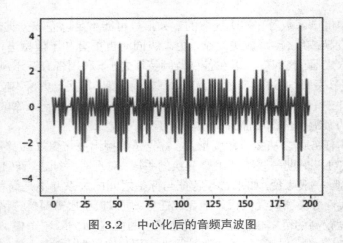

图 3.2　中心化后的音频声波图

- 电子音乐 (Electronic Music) 是使用电传簧风琴、汉门式电风琴与电吉他等电子乐器以及特雷门琴、声音合成器与电脑等电子音乐技术来制作的音乐.
- 蓝调 (Blues) 是一种基于五声音阶的声乐和乐器音乐, 起源于过去黑人的灵魂乐、赞美歌、劳动歌曲、叫喊等, 非常重视自我情感的宣泄和原创性或即兴性.
- 爵士音乐 (Jazz) 由蓝调和拉格泰姆 (Ragtime) 发展而来, 也讲究即兴, 以具有摇摆特点的 Shuffle 节奏为基础, 是非洲黑人文化和欧洲白人文化的结合.
- 摇滚音乐 (Rock) 起源于 20 世纪 40 年代末期的美国, 20 世纪 50 年代早期开始流行, 并迅速风靡全球, 它的主要特点是灵活大胆的表现形式和富有激情的音乐节奏.
- 流行音乐 (Popular) 又称商品音乐, 指的是那些以盈利为主要目的创作的兼具市场性和娱乐性并被大众所接受和喜爱的音乐.
- 世界音乐 (World Music) 泛指世界上所有的民族音乐, 这里指的是非英美及西方民歌、流行曲的音乐, 通常指发展中地区与西方音乐混和了风格的、改良了的传统音乐.

考虑到部分音乐风格之间高度的相似性和重合性, 我们将爵士音乐与蓝调合并。将摇滚音乐与流行音乐合并。之所以这么处理, 是因为爵士音乐本就传承自蓝调, 二者之间难分彼此; 而大量的摇滚音乐本就是流行音乐, 它们也不是互斥的关系.

我们知道, 完成一个分类任务必须需要特征, 也就是自变量. Fu et al. (2011) 将音频特征分成两个水平, 低水平的 (Low-Level) 和中等水平 (Mid-Level) 的, 低水平的特征还可以继续分成音色特征 (Timbre Feature) 和时序特征 (Temporal Feature). 音色特征可以捕捉音频数据中的音色, 时序特征可以捕捉声音随着时间发生的变化. 低水平的特征是通过对音频信号的傅里叶变换、频谱分析、倒频谱分析以及自回归建模等得到的. 表 3.1 列举了常见的几种低水平特征.

- 过零率 (ZCR) 指的是每帧信号内, 信号过零点的次数, 体现的是信号的频率特性, 一般来说, 过零率越高, 表明信号频率越高 (在计算过零率之前, 信号需要中心化).
- 短时能量 (Energy) 是指每帧信号的平方和, 体现的是信号能量的强弱.
- 能量熵 (Energy Entropy) 描述的是信号的时域分布情况, 体现的是信号的连续性.
- 频谱中心 (Spectral Centroid) 又称频谱一阶距, 频谱中心的值越小, 表明越多的频谱能量集中在低频范围内, 如声音 (Voice) 与音乐 (Music) 相比, 通常频谱中心较低.

- 频谱延展度 (Spectral Spread) 又称频谱二阶中心矩, 它描述了信号在频谱中心周围的分布状况.
- 谱熵 (Spectral Entropy), 根据熵的特性可以知道, 分布越均匀, 熵越大. 能量熵反映了每一帧信号的均匀程度, 如说话人频谱由于共振峰存在显得不均匀, 而白噪声的频谱更加均匀.
- 频谱通量 (Spectral Flux) 描述的是相邻帧频谱的变化情况.
- 频谱滚降点 (Spectral Rolloff) 指的是频谱中频率成分强度的 0.85 分位点.
- 梅尔频率倒谱系数 (MFCC) 是一种十分重要的音频特征, 它先对语音进行分帧和加窗操作, 对每一个短时分析窗, 通过快速傅里叶变换 (FFT) 得到其频谱, 然后利用 Mel 滤波器组对频谱进行滤波得到 Mel 频谱, 最后在 Mel 频谱上进行倒谱分析获得梅尔频率倒谱系数, 这个系数就是这帧语音的特征, 一般取第 1 个到第 13 个系数作为 MFCC 值.

表 3.1　常见的部分音频特征

类别	特征
音色	过零率
	短时能量
	能量熵
	频谱中心
	频谱延展度
	谱熵
	频谱通量
	频谱滚降点
	梅尔频率倒谱系数
时序	Statistical Moments (SM)
	Amplitude Modulation (AM)
	Auto-Regressive Modeling (ARM)

3.2.2　混合动力模型架构

音乐风格识别是一个并不容易的非线性任务, 仅仅使用低水平的音频特征对于分类任务来说并不足够. 因此, 我们需要对原始音频特征进行适当的加工, 构造出更多的有助于分类任务的新特征, 这样的过程一般叫特征工程, 即通过领域知识、特征分箱、简单的函数变换或者不同特征之间的交叉组合, 来挖掘甚至生产出大量的特征, 再通过一定的初筛手段或者直接用模型来挑选那些真正有效的特征, 最终用于各种各样的分类任务. 甚至有人总结道:"特征决定了最终预测效果的上限, 而模型只是来逼近这个上限." 大量的业界经验也表明, 通过种种手段加工出与预测目标强相关的少量特征, 给预测精度带来的提升, 往往远超过更换模型或者细致的模型调参.

生产更多特征 (自变量) 主要有三种方式: 第一种是从业务需求、领域知识出发, 结合现有原始数据, 人为构造出之前没有过的特征, 这是从 0 到 1 的创新. 表 3.1 中的所有音频特征均属于此类. 但构造这类特征难度很大, 有时需要精深的专业知识. 第二种是现有特征的线性和非线性变换, 比如常见的标准化、对数化处理、分箱操作、one-hot 编码、woe 编码等, 还有特征之间的交叉组合, 例如两特征相乘、相除、拼接、组合等, 这在现实中是最常用的类型. 第三种是依据现有特征和标签 (或因变量) 构建一个有监督模型, 将这个模型的输出或者模型本身的结构, 作为新的特征, 再输送给第二层有监督模型, 比如将决策树的每一个分支都作为一个 0-1 二元特征.

与上述几种方案不同, 脸书的团队 He et al. (2014) 提出了一种新的混合动力模型架构 (Hybrid Model Structure), 并将之用于广告点击率 (CTR) 预估. 它是建立了嵌套的两个模型, 首层为提升 (Boosting) 决策树模型, 次层是逻辑回归, 主要结构见图 3.3. 我们先在原始特征和训练样本上建立提升决策树模型, 然后将提升决策树模型的结构转化成新的输入, 作为逻辑回归模型的自变量, 再训练一个线性的逻辑回归模型.

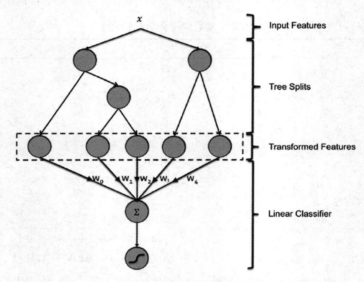

图 3.3　混合动力模型架构

He et al. (2014) 指出, 提升决策树是一种十分强大且十分方便的对原始特征进行非线性变换的方法. 只需将提升决策树的每个基学习器 (即每一棵决策树) 作为一个离散型特征, 其取值即为决策树各个叶子节点的下标, 然后对这个离散型特征使用 one-hot 编码, 形成多个 0-1 二元特征. 例如, 我们考虑图 3.3 中包含两棵子树的梯度决策树模型, 第一棵决策树有 3 个叶节点, 第二棵决策树有 2 个子节点, 如果一个样本在第一棵子树中落入了第二个叶节点中, 在第二棵子树中落入了第一个叶节点中, 那么这个样本的原始特征就被转化成一个二元向量 [0, 1, 0, 1, 0], 这个二元向量将作为新的自变量, 用于建立第二层的线性逻辑回归模型. 事实上, 我们相当于将训练出来的提升决策树模型的树结构本身, 解析成了取值 0-1 的向量, 这个向量的每个元素都依次与提升决策树的叶节点对应, 取值为 0 表示样本不会落在这个叶节点中, 取值为 1 则表明会落在其中. He et al.(2014) 中提升决策树模型采用的是梯度提升机 (Gradient Boosting Machine), 在每一轮学习中, 都会基于前一轮的残差训练一棵新的树. 我们可以将这种基于提升决策树的特征变换, 理解成将实值向量转化为 0-1 二元向量的有监督特征编码方案. 决策树每一条从根节点到叶节点的路径, 都对应了一个规则 (这条规则涉及多个特征及阈值), 梯度提升机有多少个叶节点, 就能产出多少条规则. 第二层在新产出的 0-1 二元特征上训练一个线性的逻辑回归模型, 其实就相当于在诸规则组成的规则集上, 学习每一条规则所应有的权重.

对于本案例音乐风格识别的多分类任务来说, 我们也借鉴此混合动力架构, 但在首层模型的选取上略有不同. 输入特征仍是表 3.1 中提到的那些低水平音色类音频特征, 然后在诸音频特征的基础上训练一个损失敏感型的多分类 Boosting 模型, 接着将此 Boosting 模型的每棵决策树的结构转化为多个 0-1 二分类变量 (变量数目与所有基分类器的叶节点数目相同), 最后在这些二分类变量上训练施加带 L1 惩罚的 one-vs-rest 策略下的多分类逻辑回归模型. 这样的混合动力模型架构有如下优点:

- 表 3.1 中的低水平音色音频特征虽然刻画了音频信号的诸多特性, 但是直接用于音乐风格识别任务仍显得不足, 树模型可以对原始特征进行十分有效的非线性组合, 形成强有力的规则集, 这些新挖掘的规则集可以作为新的特征, 用于之后的分类任务.
- 实验结果显示, 混合动力模型显著优于逻辑回归线性模型, 也明显优于单独的 Boosting 模型, 这表明嵌套的模型结构的确改善了预测效果.
- 逻辑回归的系数估计可以帮助我们了解不同规则的重要性, 某条规则重要性高意味着原始特征的某种非线性交叉效应对于分类任务有很好的预测能力.
- 训练一个合适的 Boosting 模型的计算量非常大, 远高于训练逻辑回归模型, 因此可以低频率训练 Boosting 模型, 然后复用此 Boosting 模型的结构, 将其转化为 0-1 二值特征, 高频率训练逻辑回归模型, 这样在保证较高预测精度的同时大大降低了运算量.

3.3 描述分析

3.3.1 数据来源及简介

我们在某音乐平台上, 搜索按照音乐风格对音乐进行分类规整的歌单, 并批量下载, 得到原始数据集, 保存在目录 MusicData, 格式为 mp3. 经过简单统计, 音乐共计 709 首, 音乐风格的分布见表 3.2. 每首歌的长度为 1 ~ 6 分钟不等, mp3 音乐文件音频信号的采样频率均为 44 100HZ, 即每一秒都有 44 100 个信号点.

表 3.2 不同风格音乐文件数目

音乐风格	原始数目	切割后数目
古典音乐	107	1 180
电子音乐	96	624
爵士音乐&蓝调	240	1 399
摇滚音乐&流行音乐	166	1 235
世界音乐	100	760
共计	709	5 198

3.3.2 数据加工

大部分音乐文件的时长都在 3 分钟以上, 但是, 我们并不需要这么长的音频信号序列才能确定音乐风格, 而且越长的音频序列, 意味着更高维的问题, 需要的样本更多, 后期建模时的计算量也越大, 所以有必要截取一个子音乐片段. 这里, 我们选择将每个音乐文件首尾各 15 秒钟去除, 然后将剩下的文件等分成 30 秒钟的子片段, 每个子片段作为一个新的样本, 其音乐风格与母体音乐文件相同. 这里的子片段长度 30 秒的选取带有一定的主观性和经验性, 经过长时间的试听, 我们认为 30 秒的时间在大部分情况下可以用来对音乐风格进行推测, 同时我们获取了更多的样本, 问题的维度以及计算量也可以得到有效控制. 删除首尾的 15 秒钟, 主要是因为部分乐曲在曲首和曲尾有较长的低音调、低音响甚至是无声片段, 很显然, 这些片段并无多大用处, 会影响样本质量. 对于不是 30 秒整数倍的音乐文件, 最后不足 30 秒的文件被删除. 这部分对应的处理程序为 1_splitMusic.py, 如下所示. 切割之后不同风格的音乐文件数目见表 3.2.

```
# -*- coding: utf-8 -*-

from pydub import AudioSegment import os

#源文件夹和目标文件夹
origin_dir = "/root/music/MusicData"
dest_dir = "/root/music/MusicData30s"
music_type = ["classical", "electronic", "jazz&blues", "rock&pop", "WorldMusic"]

#建立目标子文件夹
if not os.path.exists(dest_dir):
    os.mkdir(dest_dir)
for item in music_type:
    child_dir = dest_dir + "/" + item
    if not os.path.exists(child_dir):
        os.mkdir(child_dir)

#将所有音乐文件切割成30s的片段，首尾各去除15s
#将音乐文件重命名
for item in music_type:
    child_dir = dest_dir + '/'+ item
    origin_child_dir = origin_dir + '/'+ item
    cnt = len(os.listdir('/root/music/MusicData30s/'+ item))//2 #计数变量
    for music_file in os.listdir(origin_child_dir):
        mp3 = AudioSegment.from_mp3('/'.join([origin_child_dir, music_file]))
        if mp3:
            mp3 = mp3[15*1000:(len(mp3)-15*1000)] #去除首尾各15s
        if mp3:
```

```
            piece_len = 30*1000 #每首音乐片段的长度
            num = int(len(mp3)/(piece_len))
            for i in range(num):
                mp3[i*piece_len:(i+1)*piece_len].export('/'.join([child_dir,str(cnt)+'
.mp3']), format='mp3')
                cnt += 1
    print(item + 'done.')
```

为了方便后续处理, 将 mp3 格式的数据转换为 wav 格式, 对应程序为 2_Mp3toWav.py, 如下所示.

```
from pyAudioAnalysis import audioBasicIO
#from pyAudioAnalysis import audioFeatureExtraction
import pydub
import os

mp3_dir = "/root/music/MusicData30s"
music_type = ["classical", "electronic", "jazz&blues", "rock&pop", "WorldMusic"]

for item in music_type:
    child_dir = mp3_dir + '/'+ item
    for music_file in os.listdir(child_dir):
        if music_file.endswith('.wav'):
            continue
        #print(music_file)
        music_path = child_dir + '/'+ music_file
        song = pydub.AudioSegment.from_mp3(music_path)
        song.export(music_path.replace('.mp3', '.wav'), format='wav')
    print(item + 'done.\n')
```

为了提高模型的稳健性, 同时也为了扩大样本量, 给每个音乐文件都添加了高斯分布和均匀分布的白噪声, 这样同一份音乐文件就会有三个版本: 原始版本、带高斯分布白噪声的版本以及带均匀分布白噪声的版本. 通过这样的数据增强手段, 上述样本量都扩大为原来的 3 倍, 后期切分训练集和测试集的时候, 同一首音乐的三个版本必须同时位于训练集或同时位于测试集, 以免做出过于乐观的估计. 程序 3_BatchAddWhileNoice.py 实现了上述功能.

```
# -*- coding: utf-8 -*-
# Author:   Xu Hanhui
# 此脚本用来为wav音乐文件批量添加均匀分布和高斯分布白噪声

import os
import wave
import librosa
import numpy as np
```

```python
import copy
def AddNoise(data, MusicType='Normal'):
    DataNoise = copy.deepcopy(data)
    if MusicType == 'Uniform':
        wn = np.random.uniform(-1,1,data.shape[1])
    else:
        wn = np.random.normal(0,1,data.shape[1])
    DataNoise[0,] = 0.99*data[0,] + 0.01*wn
    if MusicType == 'Uniform':
        wn = np.random.uniform(-1,1,data.shape[1])
    else:
        wn = np.random.normal(0,1,data.shape[1])
    DataNoise[1,] = 0.99*data[1,] + 0.01*wn
    return DataNoise

AllDataPath = '/root/music/MusicData30s'
MusicGenre = ["classical", "electronic", "jazz&blues", "rock&pop", "WorldMusic"]

for item in MusicGenre:
    ChildPath = AllDataPath + '/'+ item
    for MusicFile in os.listdir(ChildPath):
        if MusicFile.endswith('.mp3') or MusicFile.endswith('Noise.wav'):
            continue
        MusicPath = ChildPath + '/'+ MusicFile
        data, fs = librosa.core.load(MusicPath,sr=44100, mono=False)
        for WhiteNoiseType in ['Normal', 'Uniform']:
            DataNoise = AddNoise(data, MusicType=WhiteNoiseType)
            PathNoise = MusicPath.replace('.wav', WhiteNoiseType+'Noise.wav')
            librosa.output.write_wav(PathNoise, DataNoise, fs)
    print(item + 'done.\n')
```

3.3.3 音频特征提取

经过 3.3.2 节的处理, 我们得到了 $5\,198 \times 3$ 个音乐样本, 每个样本时长 30 秒, 共有 5 种音乐类别. 由于每个音乐文件的音频采样频率都为 44 100HZ, 即每一秒都有 44 100 个信号点, 故 30 秒的音频文件会包含 1 323 000 个采样点. 如果每个样本都要展开成如此之长的音频序列, 那么会带来庞大的计算量, 而且信息明显冗余, 很多的计算是浪费的. 为了降低计算量, 提高效率, 让问题的复杂度变得可控, 有必要采取一定措施对长音频序列所蕴含的信息进行浓缩和萃取. 也就是说, 在构建最终的分类模型之前, 首先要做的就是音频特征的提取 (Feature Extraction).

幸运的是, 在信号处理领域, 前人已经从音频信号和其声谱图中开发出了很多成熟可用的音频特征，比如过零率和梅尔频率倒谱系数等。程序 4_FeatureExtraction.py 进行音频特征提取.

```python
import os
import numpy as np
import pandas as pd

from pyAudioAnalysis import audioBasicIO
from pyAudioAnalysis import ShortTermFeatures
import pydub

WavPath = '/root/music/MusicData30s'
MusicType = ["classical", "electronic", "jazz&blues", "rock&pop", "WorldMusic"]

#计数变量
cnt = 1

for item in MusicType:
    ChildPath = WavPath + '/'+ item
    FileNames = [wav for wav in os.listdir(ChildPath) if (wav.endswith('.wav') and not wav.endswith('Noise.wav'))]
    ntrain = int(len(FileNames)*0.8)
    for CV in ['Train', 'Test']:
        if CV == 'Train':
            start, end = 0, ntrain
        else:
            start, end = ntrain, len(FileNames)
        for OriginWav in FileNames[start:end]:
            for wav in [OriginWav, OriginWav.replace('.wav','UniformNoise.wav'), OriginWav.replace('.wav','NormalNoise.wav')]:
                SongPath = ChildPath + '/'+ wav
                [Fs, x] = audioBasicIO.read_audio_file(SongPath)
                #双声道取平均，变成单声道
                new_x = (x[:,0] + x[:,1])/2
                #这里的win和step参数要是不为整数，会被强制变成整数
                F, name = ShortTermFeatures.feature_extraction(new_x, Fs, 0.50*Fs, 0.25*Fs, deltas = False)
                #将每首音乐的特征值拉平，并堆叠
                if cnt == 1:
                    feats = np.reshape(F,(1,-1))
                else:
                    feats = np.concatenate((feats,np.reshape(F, (1,-1))), axis = 0)
                cnt += 1
```

```
        cnt = 1
        feats = pd.DataFrame(feats)
        cnames = []
        for AudioFeat in name:
            for i in range(int(feats.shape[1]/len(name))):
                cnames.append(AudioFeat + '_'+ str(i+1))
            feats.columns = cnames
            feats.to_csv(WavPath+'/'+item+CV+'.csv', header=True, sep=',', index=False)
        print(CV + 'done.')
    print(item + 'done.')
file= open('/root/music/MusicData30s/colnames.txt', 'w')
for n in cnames:
    file.write(str(n))
    file.write('\n')
file.close()
```

提取音频特征最朴素的方式,莫过于对音频序列进行分箱(切割),在每个箱子(子序列)内计算一些简单统计量或者音频特性(如过零率、能量等),然后用这些统计量来作为此子序列的代表,原始的子序列就可以抛弃掉,只保留浓缩之后的音频特征. 也就是说, 分箱操作产生了多少个箱子, 就有多少批音频特征, 将所有这些统计量连接起来, 就形成了新的音频特征向量, 其长度为 mbin×nfeat (mbin 为分箱数目, nfeat 为每个箱子内提取的音频特征种类数). 最终我们用长度为 mbin×nfeat 的音频特征向量, 代替原始的长度为 1 323 000 的音频序列, 作为自变量, 参与对音乐风格的识别. 关于分箱操作, 有一点需要注意, 常见的分箱操作产生的箱子是相邻但是互不相交的, 对于音频信号, 我们并不采取这种方案, 考虑到音频信号的连续性, 允许相邻的箱子有重合, 重合度一般为箱子长度的 1/2. 也就是说, 如果信号点的数目为 SignalLen, 分箱宽度为 BinLen, 相邻分箱重合宽度为 BinLen/2, 那么最终产生的箱子数目为 $\frac{\text{SignalLen} - \text{BinLen}/2}{\text{BinLen}/2}$. 本例中, 分箱宽度设置为 0.5×44100, 即 0.5 秒, 相邻分箱重合宽度设置为 $0.25 \times 44\,100$, 即 0.25 秒. 之后, 随机切割训练集和测试集, 分别占比 80% 和 20%, 即训练样本 $4\,158 \times 3$ 个, 测试样本 $1\,040 \times 3$ 个 (之所以是 3 倍是因为 3.2 节中利用白噪声做了数据增强).

3.4　混合动力模型架构

3.4.1　两个基础模型的预测效果

在应用混合动力模型之前, 首先看一下两个基础模型的预测效果. 对于我们研究的音乐风格识别任务, 一共有 5 个类别: Classical, Electronic, Jazz&Blues, Rock&Pop, WorldMusic. 之前的程序对 5 个类别分割处理, 程序 5_MixData.sh 将数据合并在一起.

```
#删除一些文件
rm -rf TmpTrain.csv TmpTest.csv AllTrain.csv AllTest.csv
#表头
cat Header >> AllTrain.csv
cat Header >> AllTest.csv
#混合各种音乐风格的文件,并加上标签
for MusicType in 'classical''electronic''jazz&blues''rock&pop''WorldMusic'
do
    echo -e $MusicType
    awk -F"," '{if(NR>1) print $0",'$MusicType'"}'$MusicType"Train.csv" >> TmpTrain.csv
    awk -F"," '{if(NR>1) print $0",'$MusicType'"}'$MusicType"Test.csv" >> TmpTest.csv
done
echo -e "\n"

#统计label
awk -F"," '{print $NF}'TmpTrain.csv | uniq -c
awk -F"," '{print $NF}'TmpTest.csv | uniq -c
echo -e "\n"

#随即打乱并再次统计label
shuf TmpTrain.csv >> AllTrain.csv
shuf TmpTest.csv >> AllTest.csv
echo -e "\n"
rm -rf TmpTrain.csv TmpTest.csv
```

音乐样本仍是表 3.2 中的 $5\,198 \times 3$ 个样本. 原始低水平音色音频特征为前两节中的设置, 30 秒的音频样本被切割成相邻重叠的 119 个分箱, 每个分箱提取 21 种音频特征 (其中包含 MFCC 的 13 个系数), 从而形成长度为 2 499 的特征向量 (即自变量). 通过下面的程序 6_SingleFeatureAnalysis.py, 我们在训练集上训练 one-vs-rest 策略下的多分类逻辑回归模型, 然后在测试集上进行测试, 得到的混淆矩阵如表 3.3 所示, 可以算出预测精度为 0.702. 很明显, 线性分类器对音乐风格的分类精度并不够高, 有较多的 WorldMusic 被错判为 Jazz&Blues, Rock&Pop 被错判为 Jazz&Blues, Jazz&Blues 被错判为 Rock&Pop. 为了进一步提高分类性能, 还需要考虑更复杂的分类器, 或者挖掘更有效的音频特征.

```
#删除一些文件
rm -rf TmpTrain.csv TmpTest.csv AllTrain.csv AllTest.csv
#表头
cat Header >> AllTrain.csv
cat Header >> AllTest.csv

#混合各种音乐风格的文件,并加上标签
for MusicType in 'classical''electronic''jazz&blues''rock&pop''WorldMusic'
```

```
do
    echo -e $MusicType
    awk -F"," '{if(NR>1) print $0",'$MusicType'"}'$MusicType"Train.csv" >> TmpTrain.csv
    awk -F"," '{if(NR>1) print $0",'$MusicType'"}'$MusicType"Test.csv" >> TmpTest.csv
done
echo -e "\n"

#统计label
awk -F"," '{print $NF}'TmpTrain.csv | uniq -c
awk -F"," '{print $NF}'TmpTest.csv | uniq -c
echo -e "\n"

#随即打乱并再次统计label
shuf TmpTrain.csv >> AllTrain.csv
shuf TmpTest.csv >> AllTest.csv
echo -e "\n"
rm -rf TmpTrain.csv TmpTest.csv
```

表 3.3 LR 预测结果的混淆矩阵

	Classical	Electronic	Jazz&Blues	Rock&Pop	WorldMusic
Classical	650	0	19	20	19
Electronic	2	279	31	44	19
Jazz&Blues	52	32	510	136	110
Rock&Pop	28	35	123	503	52
WorldMusic	27	22	104	54	249

决策树是一种十分典型的非线性分类器, 它的优点是偏差小, 可以完成对特征的复杂非线性组合, 并考察不同特征之间的低阶甚至高阶交叉效应对因变量的影响, 它的缺点是方差大, 十分容易过拟合, 需要配合一定的剪枝技巧. 单个学习器的学习和预测性能终归有限, Freund 和 Schapire(1997) 提出了一种集成方法 (Boosting), 用于将多个基学习器的预测结果综合起来, 进而完成最终的预测. 具体地, Boosting 首先在等权重的样本上训练一棵基分类器, 再根据此基学习器在现有样本上的预测结果, 调整每个样本的权重, 对于分类错误的样本, Boosting 增大其权重; 对于分类正确的样本, Boosting 降低其权重, 然后在新调整权重的样本上再次训练基学习器, 如此串行迭代, 直到达到事先设置好的基学习器数目为止. Boosting 中的基学习器大都采用较矮的决策树, 比如深度为 1 或者 2 的分类与回归树 (CART), 而基学习器的数目一般多达数百个, 最终的预测结果就是这数百个基学习器预测结果的加权平均或者加权投票. Boosting 有很多变体, Boosting 族的学习器一般都能达到较好的预测效果, 每年的 Kaggle 数据竞赛的前几名, 都会在其模型中采用一定的集成策略. 因此这里也将集成策略用于音乐风格识别任务.

接下来使用损失敏感型提升算法分析数据, 对应程序 7_MultiBoost5classes.py 如下所示.

```r
#此程序实现了cost-sensitive multi-class boosting
library(parallel)

#kvec是类别向量，K是类别数,类别标记必须是1,2,...,K
#输出len(kvec)*K矩阵,对角线是1，其余元素均为-1/(K-1)
#使用到这个函数的时候,kvec是训练集或测试集每个样本的预测结果
trsf=function(kvec, K) {
    res=matrix(-1/(K-1),length(kvec),K)
    #for ( i in 1:length(kvec) ) res[i,kvec[i]]=1
    res[cbind(1:length(kvec), kvec[1:length(kvec)])] <- 1
    return(res)
}

#输出len(Kvec)*K矩阵，对角线是1，其余元素均为-1
trsf1=function(kvec, K) {
    res=matrix(-1,length(kvec),K)
    #for ( i in 1:length(kvec) ) res[i,kvec[i]]=1
    res[cbind(1:length(kvec), kvec[1:length(kvec)])] <- 1
    return(res)
}

#输出len(kvec)*K矩阵，对角线是1，其余元素均为0
trsf2=function(kvec, K) {
    res=matrix(0,length(kvec),K)
    #for ( i in 1:length(kvec) ) res[i,kvec[i]]=1
    res[cbind(1:length(kvec), kvec[1:length(kvec)])] <- 1
    return(res)
}

#将一个向量标准化
std<-function(x) { return((x-mean(x))/sd(x)) }
#求一个向量最大元素的位置
max.ind=function(x) return(order(-x)[1])

#fn为函数，lower和upper为上下界，n.try为一维网格的数目
#一维网格搜索函数的最小点和最小值
my.optimize=function(fn,lower,upper,n.try=20) {
    xseq=seq(lower,upper,length=n.try)
    #res=rep(0,n.try)
    #for ( i in 1:n.try) res[i]=fn(xseq[i])
    res <- sapply(xseq[1:n.try], fn)
    #min.ind=order(res)[1] #寻找最小值及其位置完全不用对所有元素进行排序，否则太慢了
```

```r
      min.ind=which.min(res)
      return(list(minimum=xseq[min.ind],objective=res[min.ind]))
}

#xlearn和xtest分别为当前待分裂节点中包含的训练和测试样本
#依照某种损失准则(即mat矩阵),寻找xlearn的最优切割列和最优切割点,并记录一些信息
split.m=function(xlearn, ylearn, xtest, mat, colsample=1.0) {
      n=nrow(xlearn)
      n.test=nrow(xtest)
      p=ncol(xlearn)
      K=ncol(mat)

      #此函数计算第j个变量的最佳切割点及对应的损失
      find.best.split = function(j){
         #split.info[1]记录第j个变量的最佳切割点,[2]记录对应的损失,[3]记录变量名
         split.info=c(1,0,j)
         #挑出xlearn的第j列大于aa的那些行
         #对于mat矩阵的这些行,求其列和最小值
         f.foo=function(aa) {
            aa.ind=which(xlearn[,j]>=aa)
            mat.ind=mat[aa.ind,]  #特征取值大于等于阈值aa的那些行
            if (!is.matrix(mat.ind)) mat.ind=t(mat.ind)
            mat.ind.n=mat[-aa.ind,]  #特征取值小于阈值aa的那些行
            if (!is.matrix(mat.ind.n)) mat.ind.n=t(mat.ind.n)
            #按当前特征和阈值分割后左右子节点的损失和
            return( min(apply(mat.ind,2,sum))+min(apply(mat.ind.n,2,sum)) )
         }
         #如果xlearn的第j列存在一定大小的波动,那么使用网格搜索
         #找到最优切割点,使得f.foo函数值最小,并记录下来
         if (sd(xlearn[,j])>10^(-3)) {
            foo=my.optimize(f.foo,min(xlearn[,j]),max(xlearn[,j]))
            split.info[1]=foo$minimum
            split.info[2]=foo$objective
         } else {
            #如果xlearn的第j列值几乎相同,就将最小值定成一个很大的值
            split.info[1]=0
            split.info[2]=10^6
         }
         return(split.info)
      }
      #res记录每一个变量的最佳分割点及对应的损失
      #res=t(sapply(sample(1:p, size=as.integer(p*colsample)), find.best.split))
```

```r
    #开始并行化
    CoreNum <- detectCores() #- 1
    MyCluster <- makeCluster(CoreNum, type = 'FORK')
    res = t(parSapply(MyCluster, sample(1:p, size=as.integer(p*colsample)),
find.best.split))
    stopCluster(MyCluster)

    #切割时的最优列
    ind.best = which.min(res[,2])
    j.best=res[ind.best,3]
    #对于xlearn和xtest的最优切割列,取值大于最优切割点,则样本标记为2,否则标记为1
    #这么标记是为了在learner.m函数中根据样本与切割点的关系将其分配到左右子节点中
    learn.ind=as.numeric(xlearn[,j.best]>=res[ind.best,1])+1
    test.ind=as.numeric(xtest[,j.best]>=res[ind.best,1])+1

    return(list(learn.ind=learn.ind, test.ind=test.ind,
            best.var=j.best, best.threshold=res[ind.best,1]))
}

#xlearn和ylearn为训练集特征及label,xtest为测试集特征
#wgt为每个样本的初始权重
#mat为n*K,记录每个样本被分到不同类别的错判损失
#wgt和mat不能均为空,如果设置了wgt,就根据wgt设置mat,否则采用用户定义的mat
#adaboost.my调用learner.m的时候是直接设置好了mat,没有设置wgt
#level为每个基分类器(子树)的深度,这里受计算复杂度的限制,至多设置2
learner.m=function(xlearn, ylearn, xtest, wgt=NULL, mat=NULL, level=1, colsample=1.0)
{
    n=nrow(xlearn)
    n.test=nrow(xtest)
    p=ncol(xlearn)
    K=length(table(ylearn)) #类别数
    if (is.vector(wgt)) {
        #mat是n*K的损失矩阵,第i行代表第i个样本被判到K个类别的损失
        mat=matrix(rep(wgt/(K-1),K),ncol=K)
        mat[cbind(1:n,ylearn)]=0
    }

    #这两个subset初始化全为1,后续借助这俩个subset中的值来跟踪每个样本归属于哪个叶节点
    #每一层叶节点标号都是从1:2^(l-1)
    learn.subset=rep(1,n)
    test.subset=rep(1,n.test)
```

```
#记录基分类器的最优切割点
base.learner.info = c()
#基分类器是一棵深度为1或者2的二叉树
#下面的循环用于寻找每个节点的最佳特征和最佳分裂点
#分裂时选择的准则为样本总损失最小化,使用线性网格搜索寻找最优特征和最优点
for ( l in 1:level ) {
        learn.foo=learn.subset
        test.foo=test.subset
        #树的第1-1层到第1层会有2^(l-1)待分裂处
        #受限于指数复杂度,level的取值不宜过高,当然boosting每个基分类器一般取深度为1或2的树就足够了
        for ( ll in 1:(2^(l-1)) ) {
                learn.flag = (learn.subset==ll)
                test.flag = (test.subset==ll)
                if (sum(learn.flag)>1) {
                        split.foo=split.m(xlearn[learn.flag,],ylearn[learn.flag],xtest[test.flag,],mat[learn.flag,],colsample=colsample)
                        base.learner.info = c(base.learner.info, c(split.foo$best.var, split.foo$best.threshold))
                        learn.tmp=split.foo$learn.ind #标记每一行是大于分割点还是小于分割点
                        test.tmp=split.foo$test.ind
                } else {
                        #如果当前待分裂节点只包含一个样本,则全部放入左子节点中,右子节点为空
                        #当前分裂节点可能包含不止一个测试样本,但仍强制全部归入一个节点中
                        base.learner.info = c(base.learner.info, c(0,0))
                        learn.tmp=1
                        test.tmp=rep(1,sum(test.flag))
                }
                #2*(ll-1)+1和2*(ll-1)+2为当前待分裂节点的左右节点标记
                learn.foo[learn.flag]=2*(ll-1)+learn.tmp
                test.foo[test.flag]=2*(ll-1)+test.tmp
        }
        learn.subset=learn.foo
        test.subset=test.foo
}
#将基分类器的信息写出
fl <- file('base.learner.info.csv', 'a')
writeLines(paste(base.learner.info, collapse=','), fl)
close(fl)

ylearn.pred=rep(0,n)
ytest.pred=rep(0,n.test)
```

```
                for ( lll in 1:(2^level) ) {
                        learn.flag = (learn.subset==lll)
                        if (sum(learn.flag)==0) next #next语句相当于continue
                        #训练样本和测试样本的y的预测值
                        #对每一个叶节点,选取一个类别使得损失最小,此类别则是此叶节点所有样本的
预测值
                        pred.tmp=ifelse(sum(learn.flag)==1,ylearn[learn.flag],which.min(apply
(mat[learn.flag,],2,sum)))
                        ylearn.pred[learn.flag]=pred.tmp
                        ytest.pred[test.subset==lll]=pred.tmp
                }

                #返回当前子树在训练集和测试集上的预测结果
                return(list(learn=ylearn.pred,test=ytest.pred,base.learner.info=base.learner.
info))
}

#此boosting方法的每棵基分类器都是树桩
#cost为K*K的损失矩阵,记录不同类别间的错判损失
adaboost.my <- function(x.train, y.train, x.test, y.test, cost=NULL, mfinal=200,
level=1, colsample=1.0) {
    ## Initialization
    n=nrow(x.train)
    p=ncol(x.train)
    K=length(unique(y.train))
    n.test=nrow(x.test)
    #如果未指定损失矩阵,则设置为错判损失相等的损失矩阵
    if (is.matrix(cost)) c.mat=cost else c.mat=1-diag(K)

    #boosting在训练集和测试集上的累积预测结果
    #此处是初始化,之后会随着迭代不断更新
    if(file.exists('Flearn.csv')){
        Flearn <- as.matrix(read.csv('Flearn.csv', header=F, sep=','))
        colnames(Flearn) <- c(1:K)
    }else{
        Flearn=matrix(0,n,K)
    }
    if(file.exists('Ftest.csv')){
        Ftest <- as.matrix(read.csv('Ftest.csv', header=F, sep=','))
        colnames(Ftest) <- c(1:K)
    }else{
```

```
        Ftest=matrix(0,n.test,K)
}

#初始化,a.mat为n*K的矩阵,第i行为第i个样本yi被判到不同类别的损失
if(file.exists('a.mat.csv')){
    a.mat <- as.matrix(read.csv('a.mat.csv', header=F, sep=','))
    colnames(a.mat) <- c(1:K)
}else{
    a.mat=matrix(0,n,K)
    a.mat[1:n,]=c.mat[y.train[1:n],]
}

## Boosting Iterations

#记录每一轮的预测精度
acc=matrix(0,mfinal,2)
#记录每一轮的训练和测试误差
err=matrix(0,mfinal,2)
#记录每一个基分类器的信息,第一列是变量标记,第二列是分割阈值
ntrees.info = matrix(0, mfinal, 2*(2**level-1))

for (m in 1:mfinal) {
    ## Fitting the Tree
    #level为树的深度,level=1表示每棵子树是树桩
    update=learner.m(x.train, y.train, x.test, mat=a.mat, level=level,
colsample=colsample)
    ntrees.info[m,] = update$base.learner.info
    #根据每棵子树的预测结果计算当前更新的子函数
    flearn=trsf(update$learn,K)
    ftest=trsf(update$test,K)
    #当前子树在训练集上的预测损失,此处没有归一化
    f.err=sum(a.mat[cbind(1:n,update$learn)])
    all.loss=sum(a.mat)

    ## Updating
    if (f.err>0) {
        ## f.foo=function(aa) return(sum(a.mat*exp(aa*flearn)))
        ## f.coef=my.optimize(f.foo,0,100)$minimum ### Line Search
        #这里之所以是sum(a.mat)而不是论文14式中的1,是因为f.err没有做论文15式
中的归一化
        f.coef=(K-1)/K*(log((all.loss-f.err)/f.err)-log(K-1))
        #每个子分类器的权重
```

```
            ## f.coef=1/n

            Flearn=Flearn + f.coef*flearn
            Ftest=Ftest + f.coef*ftest
            a.mat=a.mat*exp(f.coef*flearn) #更新损失矩阵,这相当于调整样本的权重
            all.loss=sum(a.mat)
            if (all.loss==0) break
            a.mat=pmin(a.mat/all.loss,1e24)  #归一化
        }
        if (f.err==0) {
            Flearn=Flearn + flearn
            Ftest=Ftest + ftest
        }
        #将Flearn, Ftest, a.mat及时写出，方便程序中断时及时恢复进度
        write.table(Flearn, 'Flearn.csv', col.names=F, row.names=F, quote=F, sep=',')
        write.table(Ftest, 'Ftest.csv', col.names=F, row.names=F, quote=F, sep=',')
        write.table(a.mat, 'a.mat.csv', col.names=F, row.names=F, quote=F, sep=',')

        #记录截止到当前轮的训练和测试集上的损失
        #max.ind为自定义函数
        #apply(Flearn,1,which.max)为每个样本当前概率最大的类别
        pred.train = apply(Flearn,1,which.max)
        pred.test  = apply(Ftest,1,which.max)
        err[m,1]=sum(c.mat[cbind(y.train,pred.train)])/n
        err[m,2]=sum(c.mat[cbind(y.test,pred.test)])/n.test
        confusion.mat.train = table(y.train, pred.train)
        confusion.mat.test  = table(y.test, pred.test)
        acc[m,1] = sum(diag(confusion.mat.train))/n
        acc[m,2] = sum(diag(confusion.mat.test))/n.test
        #将err和acc及时写入文件
        fl <- file('err.csv', 'a')
        writeLines(paste(err[m,],collapse=','), fl)
        close(fl)
        fl <- file('acc.csv', 'a')
        writeLines(paste(acc[m,],collapse=','), fl)
        close(fl)

        #打印日志
        cat("***",m,"轮***\n")
        cat("train混淆矩阵：  \n")
        print(confusion.mat.train)
        print("训练精度：")
```

```r
            print(sum(diag(confusion.mat.train))/n)
            cat("test混淆矩阵:  \n")
            print(confusion.mat.test)
            print("测试精度: ")
            print(sum(diag(confusion.mat.test))/n.test)
            cat("\n")
    }
    
    #返回最终的训练集和测试集预测结果及预测损失
    return(list(ytr=pred.train, yte=pred.test, error=err,
                acc=acc, ntrees.info=ntrees.info))
}

main <- function(){
  #x.train, y.train, x.test, y.test, cost=NULL, mfinal = 200
  print(date())
  TrainData <- read.csv('../AllData30s/AllTrain.csv', header=T, sep=',')
  p <- ncol(TrainData)
  p.use = min(2499,p-1)
  n.train <- nrow(TrainData)
  music.type = levels(TrainData[,p])
  for(i in 1:length(music.type)){
    cat(i,":",music.type[i],'\n')
  }
  TrainData[,p] <- as.integer(TrainData[,p])
  x.train <- TrainData[,1:p.use]
  y.train <- TrainData[,p]
  y.train <- matrix(y.train, length(y.train), 1)
  rm(TrainData)
  
  TestData <- read.csv('../AllData30s/AllTest.csv', header=T, sep=',')
  TestData[,p] <- as.integer(TestData[,p])
  n.test = nrow(TestData)
  row.names(TestData) <- as.character((n.train+1):(n.train+n.test))
  x.test <- TestData[,1:p.use]
  y.test <- TestData[,p]
  y.test <- matrix(y.test, length(y.test), 1)
  rm(TestData)
  
  print(date())
  
  mfinal <- 400
```

```r
    #根据上次中断的进度来重新更新循环次数
    if(file.exists('acc.csv')){
        AccTmp <- read.csv('acc.csv', header=F, sep=',')
        mfinal <- mfinal - nrow(AccTmp)
    }
    cat("剩余基分类器数目: ", mfinal, "\n")
    level <- 2 #指定每棵树的深度
    #指定损失矩阵
    cost <- t(matrix(c(0,1/1180,1/1180,1/1180,1/1180,
                       1/624,0,1/624,1/624,1/624,
                       1/1399,1/1399,0,1/1399,1/1399,
                       1/1235,1/1235,1/1235,0,1/1235,
                       1/760,1/760,1/760,1/760,0), nrow = length(music.type)))
    cost <- t(matrix(c(0,1,1,1,1,
                       1,0,1,1,1,
                       2,1,0,4,2,
                       1,1,4,0,1,
                       1,1,4,1,0),nrow = length(music.type)))
    #cost<-cost*matrix(rep(c(91,144,446,669,447),length(music.type)),length(music.type))
    #cost<-NULL
    print(cost)
    #训练结果
    res <- adaboost.my(x.train, y.train, x.test, y.test, cost=NULL, mfinal = mfinal,
level=level, colsample=0.9)
    #打印每一轮的预测损失和预测精度
    cat("***error:***\n")
    print(res$error)
    cat("***acc:***\n")
    print(res$acc)
    cat("\n")
    #记录每个基分类器的信息
    ntrees.info = data.frame(res$ntrees.info)
    cnames <- c() #列名
    for (i in c(1:(2**level-1))){
        cnames<-c(cnames,c(paste("split.var",i,sep='.'),paste("split.threshold",i,sep='.')))
    }
    colnames(ntrees.info) <- cnames
    write.csv(ntrees.info, 'AdaboostModelInfo.csv', quote=F, row.names=F)
    print(date())
}
main()
```

先设置损失矩阵为等损失的, 即不同类别之间的错判损失相同, 并限制每棵基分类器 (即决策树) 的深度为 2, 基学习器的数目为 400, 训练集、测试集以及音频特征的设置均与之前相同, 最终在测试集上预测的混淆矩阵见表 3.4, 预测精度为 0.775 6.

表 3.4 等损失矩阵时多分类 Boosting 预测结果混淆矩阵

	Classical	Electronic	Jazz&Blues	Rock&Pop	WorldMusic
Classical	695	0	7	6	0
Electronic	6	303	27	32	7
Jazz&Blues	58	20	580	132	50
Rock&Pop	32	30	111	546	22
WorldMusic	31	19	87	23	296

Boosting 模型在迭代过程中, 在训练集上和测试集上的预测误差见图 3.4.

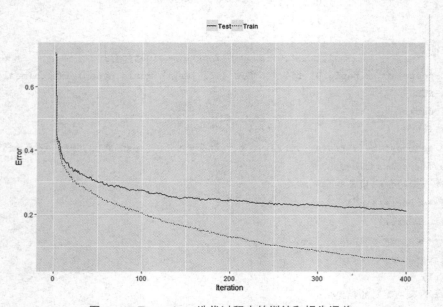

图 3.4 Boosting 迭代过程中的训练和损失误差

同时, 如果单独看每一个类别的预测情况 (二分类问题), 精确率和召回率是我们比较关心的问题, 图 3.5 展现了 5 种音乐风格各自的预测精确率和召回率.

这里的预测效果比逻辑回归模型要好, 但是部分类别预测失败的情况仍比较严重, 可以通过调整损失矩阵来平衡各个类别间的误判, 在整体预测精度不降的情况下, 使部分类别的预测准确率或召回率得到改善.

需要了解的是, 虽然不同风格的音乐之间会有所差异, 但是部分音乐风格之间存在不小的相似性. 例如, Rock&Pop 中包含的都是摇滚音乐和流行音乐, 事实上, 流行音乐是一种比较宽泛的概念, 更严格地说, 它应该叫做 "商品音乐", 指的是那些以盈利为主要目的创作的兼具市场性和娱乐性, 并被大众所接受和喜爱的音乐, 它与蓝调、爵士音乐、摇滚音乐之间

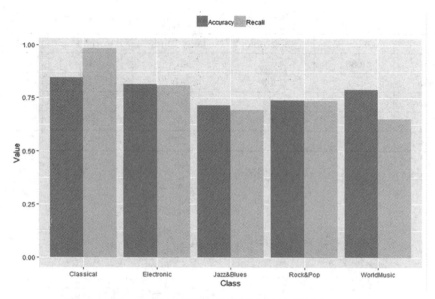

图 3.5　不同音乐风格的准确率和召回率

并不是简单的排斥关系, 而是交叉关系, 很多的摇滚乐、爵士音乐和蓝调事实上也是商品音乐, 享有大量的听众, 甚至在全球流行, 但是摇滚音乐、爵士音乐、蓝调等又有自己的独特风格, 并不都是流行音乐, 这些专属特点在很多流行音乐中未必能见到; 同时, 世界音乐指的是非英美及西方民歌、流行曲的音乐, 通常指发展中地区与西方音乐混和了风格的、改良了的传统音乐, 多来自非洲、拉丁美洲及亚洲, 这些音乐除了带有浓厚的本地传统色彩, 还混杂着一些西方现代音乐风格的味道, 比如很多世界音乐会借鉴蓝调和爵士音乐的唱法以及乐器. 这也可以从表 3.4 的混淆矩阵看出来, Jazz&Blues 和 Rock&Pop 之间发生了很多的误判, WorldMusic 也常被误判为 Jazz&Blues. 因此, 我们难以对每一首音乐做出准确且唯一的风格判定, 因为它可能同属于两种甚至多种音乐风格, 我们只能在将相似音乐风格合并到一起的同时, 再尽可能降低误判.

为了改善等损失矩阵对 Jazz&Blues、Rock&Pop 以及 WorldMusic 的部分误判情况, 可以调节损失矩阵, 使这几个音乐风格之间的误判代价更大, 进而使得分类器向这些类别倾斜, 增大它们预测的准确度. 但是, 天下无免费的午餐, 提高部分音乐风格的预测精度, 必然是以损失其他音乐风格预测精度为代价的, 所以, 我们需要在不同音乐风格的预测精度上做权衡. 应尽量避免的情况当然是某些类别的预测准确率极高, 但是其他类别的预测准确率极低, 可以适当降低高准确率类别的误判损失, 提高低准确率类别的误判损失, 使得每个类别都能达到较高的预测精度.

我们适当调整误判损失矩阵 (看一下程序中是怎样调整损失矩阵的), 再次训练损失敏感型的 Boosting 模型, 其预测混淆矩阵见表 3.5.

表 3.5　损失敏感型多分类 Boosting 预测结果的混淆矩阵

	Classical	Electronic	Jazz&Blues	Rock&Pop	WorldMusic
Classical	680	0	3	18	7
Electronic	20	296	31	19	9
Jazz&Blues	53	23	596	117	51
Rock&Pop	21	32	101	556	31
WorldMusic	26	25	79	17	309

同时，不同类别的准确率和召回率见图 3.6.

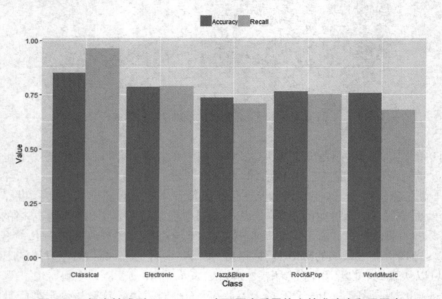

图 3.6　损失敏感型 Boosting 在不同音乐风格上的准确率和召回率

可以看到，在新的代价矩阵下，Jazz&Blues、Rock&Pop 以及 WorldMusic 的预测得到了明显改善，但是 Classical 和 Electronic 的预测效果有了略微下降，整体的预测准确度为 0.781 1，相比于错判损失相等的情况有略微提高。

3.4.2　混合动力模型架构的预测效果

树是一种十分强大的非线性转换器，它可以将原始的音频特征重新编码成一系列 0-1 取值的特征，每一个新特征都与部分原始特征的条件判断和非线性组合相对应. He et al.(2014) 创造性地将 GBDT 所有的基分类器都解析成 0-1 取值特征，并输入逻辑回归模型，使用这种混合动力模型架构提高了 CTR 预估的效果. 受此启发，在区分音乐风格这样的多分类问题下，也可以使用 Boosting 族的模型去对音频特征进行深加工，并通过解析 Boosting 的结构，得到更为高级的对于预测音乐风格有更强效力的特征，然后再套一层简单的线性分类器，最终提高分类效果. 另外，由于多分类任务常常具有误判损失不相等的情况，我们无法将所有类别都视为是平等的，所以，可以将混合动力模型架构首层的 GBDT 替换成损失敏感型的 Boosting

模型, 这样可以改善部分类别的误判情况. 对应的程序分别是 8_AdaboostOutput2LRinput.R 和 9_LRCVTuneParams.py.

```r
#将MultiBoost的诸基分类器转化为二分类特征

DataPath <- '../AllData30s/NoMetalPunkTrain.csv'
ModelPath <- 'ModelTuneInfo/ModelFile10'
NewDataPath <- '../AllData30s/TrainInputForLR.csv'

data <- read.csv(DataPath, header=T, sep=',')
n <- nrow(data)
p <- ncol(data) - 1
print("Read data over!")

ModelFile <- read.csv(ModelPath, header=T, sep=',')
n <- nrow(data)
ntree <- nrow(ModelFile)
print("Read model over!")

#每棵树4个节点,每个节点对应1个0-1变量
#不过每棵树只需3个0-1变量即可确定,剩余一个是冗余的
NewData <- data.frame(matrix(0, n, 3*ntree+1))
j <- 1
for(m in c(1:ntree)){
  NewData[,j] <- as.integer(data[,ModelFile[m,1]]<=ModelFile[m,2] &
data[,ModelFile[m,3]]<=ModelFile[m,4])
  j <- j+1
  NewData[,j] <- as.integer(data[,ModelFile[m,1]]<=ModelFile[m,2] &
data[,ModelFile[m,3]]>ModelFile[m,4])
  j <- j+1
  NewData[,j] <- as.integer(data[,ModelFile[m,1]]>ModelFile[m,2] &
data[,ModelFile[m,5]]<=ModelFile[m,6])
  j <- j+1
  NewData[,j] <- as.integer(data[,ModelFile[m,1]]>ModelFile[m,2] &
data[,ModelFile[m,5]]>ModelFile[m,6])
  j <- j+1
}
NewData[,j] <- data$label
print("Feature transformation over!")

write.csv(NewData, NewDataPath, quote=F, row.names=F)
print("Write features over!")
```

```python
# -*- coding: utf-8 -*-
#此脚本在TransformedInput的特征上建立LR模型,并对正则参数C进行调参

import numpy as np
import pandas as pd
from sklearn.linear_model import LogisticRegression
from sklearn.metrics import confusion_matrix
from sklearn.model_selection import StratifiedKFold

#读入训练数据
TrainData = pd.read_csv('TrainInputForLR.csv', sep=',', header=0)
#TrainData = pd.read_csv('TrainInputForLRTmp.csv', sep=',', header=0)

#将最后一列列名修改为'Label'
ColNames = list(TrainData.columns)
ColNames[-1] = 'Label'
TrainData.columns = ColNames

#类别去重及映射
UniqueClasses = sorted(TrainData['Label'].unique())
ClassesMap = dict(zip(UniqueClasses, range(len(UniqueClasses))))
InverseClassesMap = dict(zip(ClassesMap.values(), ClassesMap.keys()))
print(ClassesMap)
print(InverseClassesMap)
print(sorted(ClassesMap.keys()))

#将类别重新编码为数字
TrainData['Label'] = TrainData['Label'].map(ClassesMap)

#5折交叉验证
nfolds = 5
#网格调参需要遍历的列表
params = [0.1, 1.0, 5.0, 10.0, 100.0]
#网格调参
for m,C in enumerate(params):
    #记录每一折训练和测试精度的临时变量
    TmpTrainAcc = TmpTestAcc = 0
    #记录每个参数对应的5折平均训练和测试精度
    TuneParamsRecord = pd.DataFrame(np.zeros([len(params),3]))
    TuneParamsRecord.columns = ['C', 'TrainAccAvg', 'TestAccAvg']
    #初始化5折交叉验证,StratifiedKFold保证每个类别切割均匀
    skf = StratifiedKFold(n_splits=nfolds)
```

```python
#训练每一折的加L1罚的LR模型
for TrainIndex, TestIndex in skf.split(TrainData.iloc[:,:-1], TrainData['Label']):
    XTrain, XTest = TrainData.iloc[TrainIndex,:-1], TrainData.iloc[TestIndex,:-1]
    yTrain, yTest = TrainData.loc[TrainIndex,'Label'], TrainData.loc[TestIndex,'Label']
    #训练LR模型
    model = LogisticRegression(penalty='l1',dual=False,tol=0.0001,C=C,fit_intercept=True,solver='saga',multi_class='ovr',max_iter=10000)
    clf = model.fit(XTrain, yTrain)
    print(clf.n_iter_)
    #计算测试集预测精度
    PredictRes = clf.predict(XTest)
    ConfMat = confusion_matrix(y_true=yTest.map(InverseClassesMap),
                y_pred=pd.Series(PredictRes).map(InverseClassesMap),
                labels=sorted(ClassesMap.keys()))
    TestAcc = np.trace(ConfMat)/sum(sum(ConfMat))
    TmpTestAcc += TestAcc
    #计算训练集预测精度
    TrainPredictRes = clf.predict(XTrain)
    TrainConfMat = confusion_matrix(y_true=yTrain.map(InverseClassesMap),
                y_pred=pd.Series(TrainPredictRes).map(InverseClassesMap),
                labels=sorted(ClassesMap.keys()))
    TrainAcc = np.trace(TrainConfMat)/sum(sum(TrainConfMat))
    TmpTrainAcc += TrainAcc
    print(TrainAcc, TestAcc)
#记录结果
TuneParamsRecord.iloc[m,:] = [C, TmpTrainAcc/nfolds, TmpTestAcc/nfolds]
print([C, TmpTrainAcc/nfolds, TmpTestAcc/nfolds])
#将记录写出
TuneParamsRecord.to_csv('TuneParamsRecord.csv', sep='\t', header=True, index=False)
```

此外, 混合动力模型架构也可以带来计算量的优化, Boosting 族算法的调参及训练过程需要大量的计算, 但是逻辑回归的计算量要小很多, 新的混合动力模型架构可以弥补计算量上的缺点. 在有更多的数据以及计算资源时, 我们会不断更新现有模型, 以便学习到更多的特性, 未来做到更好的预测, 此时可以考虑以一个较低的频率更新首层的 Boosting 模型, 以一个较高的频率更新第二层的逻辑回归模型, 这样可以大大减小训练模型需要的计算量, 同时又能保证分类器的预测效果. 例如, 只有少量的数据更新时, 可以只更新逻辑回归模型, 当有大量的数据补充和更新时, 就可以两层模型都更新.

我们在上一小节的基础上, 解析其得到的损失敏感型 Boosting 模型的结构, 它包含 400 棵树, 每棵树的深度为 2, 即有 4 个叶节点, 这样我们可以将其转化为 800 个二分类特征 (考虑到每棵树的 4 个特征其实是有冗余的, 有 1 个 0-1 特征可以由剩余 3 个 0-1 特征线性计算得到, 因此真正有效的特征只有 600 个). 在新产出的这 600 个 0-1 特征上, 再次训练逻辑回归模型, 在测试集上预测的准确率为 0.812 7, 要优于之前的结果, 混合动力模型架构、逻辑回归和损失敏感型 Boosting 的预测准确率对比见图 3.7.

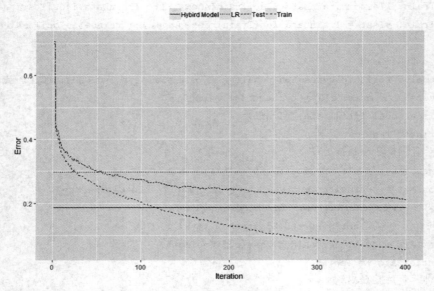

图 3.7　三种模型的预测结果的对比

可以看到, 混合动力模型架构的预测效果要明显优于逻辑回归, 也优于损失敏感型 Boosting 算法, 但是混合动力模型架构简化了原始的 2 499 个音频特征, 将其加工成 600 个新的特征, 数量上的大幅减少使得后期训练和更新逻辑回归模型时可以节省大量时间. 朴素的逻辑回归、损失敏感型的 Boosting 算法以及混合动力模型架构的第二层逻辑回归模型的训练时间对比见图 3.8 (以 400 棵树的损失敏感型 Boosting 算法的训练时间为 100), 很明显, Boosting 算法的计算成本太庞大, 在得到更高级的特征之后, 应该转而高频训练简单的逻辑回归, 降低模型迭代的计算量.

3.4.3　工程优化

1. 正则化

机器学习模型并不是越复杂越好, 复杂的模型往往会过拟合, 即对训练集拟合得非常好, 但是在测试集上的预测效果却不理想. 在训练模型时, 我们需要对模型的复杂度做一定的限制, 防止它不仅拟合了训练集中的信号, 还拟合了其中的噪声. 给定一个模型之后, 限制其复杂度的方式很多, 比如限制模型参数的范数、树的深度等. 狭义的正则化, 常指对参数向量的 L1 和 L2 范数惩罚; 广义的正则化, 可泛指一切限制当前模型复杂度、防止过拟合以提高未来预测效果的措施. XGBoost 是加强版的 GBDT, 其中有很多的正则化措施和近似算法, 用于抑制过度学习以及降低计算量, 例如, 行采样、列采样 (随机森林)、对叶节点数目以及每个叶节点代表值的 L1 和 L2 范数的惩罚、节点分裂时寻找最优切割点的近似算法 Weighted Quantile Sketch 等. 本案例在 Wang(2013) 提出的损失敏感型 Boosting 算法的基础上, 添加

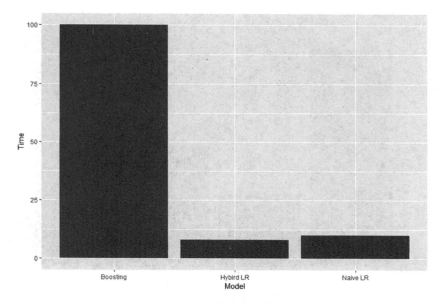

图 3.8　不同模型的计算量对比

了行采样以及列采样的正则化技巧,同时为了降低计算量,在决策树分裂寻找最优变量及其最优分割点的过程中,不再遍历所有可能的阈值,而是遍历分位点. Chen 和 Guestrin(2016) 提到,当分位点粒度较细的时候,其分割效果与精确搜索效果十分接近,但是可以大幅降低计算量,本文采用 1%～5% 分位点. 列采样是在树节点分裂过程中,不再从所有可能的候选变量中挑选,而是随机抽样,从抽取的变量中选取最佳变量及其分割阈值,这样的好处是不让某一个或者某几个变量占据绝对的话语权,从而使得每一个变量都有贡献机会,同时每棵树的结构也不会过度相似,不同的基学习器专注于学习数据背后概率分布的不同侧面,不仅降低了计算量,还起到了过拟合的作用,本文列采样的概率设置为 0.8～1.0 不等,典型取值为 0.9.

2. 并行化

Boosting 算法与随机森林 (Random Forest) 在计算上的显著差异是,后者不同基学习器的计算是完全并行的,而 Boosting 的每一棵树的生长都要依赖于前一棵树在现有训练集上的预测结果,因而算法整体上必须是串行的,但是这并不意味着不存在优化空间. 决策树每个节点分裂为两个节点的时候,都需要挑选分割变量和分割阈值,在这个过程中不同特征之间是互不影响的,所以可以实现并行,再结合列采样和分位点近似切割算法,因而可以节约大量计算时间.

我们改进了 Wang (2013) 原始算法的部分实现,主要是在决策树生长过程中利用了 CPU 多核进行并行化计算,添加 0.9 的列采样比率进而无须遍历所有特征,以及优化了一些代码的时间复杂度,图 3.9 展示了不同操作节约的计算时间,其中并行化算法是利用了一个 5 核的 CPU 进行试验的. 可以看到,近似切割算法能带来极大的计算量改进,并行化也能显著提升计算效率.

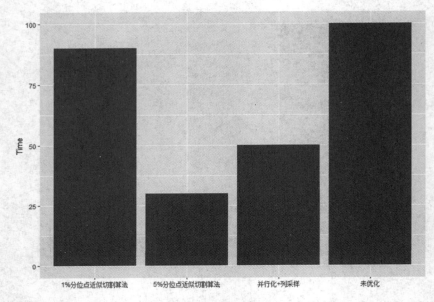

图 3.9　近似算法和并行化带来的计算量改进

3. 模型的持久化

混合动力模型架构的首层是损失敏感型 Boosting 算法, 这个模型的作用除了提高学习效果和预测精度, 还有完成原始音频特征的非线性组合. 这些新产出的特征, 除了提供给第二层的逻辑回归模型学习, 还可以输出为稳定可复用的音频特征, 用于未来逻辑回归模型的更新和其他的音频分类任务. 所以我们需要考虑将 Boosting 模型的结构以一种方便的形式保存到本地, 以便可以随时读入并恢复, 这也称作模型的持久化.

如果 Boosting 每棵树的深度都较小, 且结构完全相同, 只是每个节点的分割变量和阈值不同, 那么可以使用规整的数据框来保存信息, 并将其写入本地的逗号分隔文件或制表分割文件. 例如, 若 Boosting 的每一个基学习器都是严格的深度为 1 的树桩 (Stump), 如图 3.10 所示, 我们就可以使用形如表 3.6 的数据框存储其信息, 该数据框每一行都完整记录了基学习器的所有信息 (其中 Split.var 表示用于节点分割的变量名, Split.threshold 为其分割阈值, Weight 为基学习器的权重), 再次使用时, 只需调用事先写好的解析器即可恢复 Boosting 的结构并进行特征转换. 当然如果树的深度大于等于 2, 就需要更多的列来记录信息, 唯一需要注意的是要按照顺序存储每个节点的分割变量和阈值.

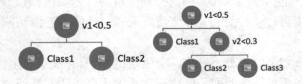

图 3.10　深度为 1 和 2 的 CART

表 3.6 深度为 1 的 CART(树桩) 的模型信息存储

Split.var	Split.threshold	Weight
v1	0.5	0.15
v2	0.3	0.18
⋮	⋮	⋮

如果每棵决策树的结构并不相同且不规整,那么可以使用 json 字符串来持久化存储模型对象. 例如, 图 3.10 右侧的决策树, 就可以使用图 3.11 中的 json 字符串来表示, 很明显, 该 json 字符串是递归的. 当我们需要存储决策树的包括叶节点概率分布在内的更多信息的时候, 只用稍微调整 json 字符串的字段和结构即可, 因而, 这是比前一种方法更为一般的存储策略. 复用时, 通过调用 json 的解析器来恢复该决策树.

```
{
    "weight":0.85,
    "structure":{
        "split.var":"v1",
        "split.threshold":0.5,
        "LeftChild":{
            "class":1
        },
        "RightChild":{
            "split.var":"v2",
            "split.threshold":0.3,
            "LeftChild":{
                "class":2
            },
            "RightChild":{
                "class":3
            }
        }
    }
}
```

图 3.11 损失敏感型 Boosting 在不同音乐风格上的准确率和召回率

> 思考:
> • 如何找到最优的代价矩阵?

第 4 章 航空数据案例分析

4.1 数据简介

该数据包括 1988—2008 年美国各机场国内航班起降记录, 每年一个文件. 变量说明见表 4.1.

表 4.1 1988—2008.csv 字段说明

变量编号	变量名	释义
1	Year	对应年份 (1988—2008)
2	Month	对应月份 (1—12)
3	DayOfMonth	航班在一个月中的哪一天起飞 (1—31)
4	DayOfWeek	航班在一星期中的哪一天起飞 (1—7)
5	DepTime	实际起飞时间 (当地时间)
6	CRSDepTime	计划起飞时间 (当地时间)
7	ArrTime	实际到达时间 (当地时间)
8	CRSArrTime	计划到达时间 (当地时间)
9	UniqueCarrier	航班所属的航空公司 (国际航空运输协会 (IATA) 航空公司代码), 对应 carriers.csv
10	FlightNum	航班号
11	TailNum	航班尾号 (飞机 id)
12	ActualElapsedTime	实际到达时间与实际起飞时间之差
13	CRSElapsedTime	预计到达时间与预计起飞时间之差
14	AirTime	空中飞行时间
15	ArrDelay	实际到达时间与预计到达时间之差
16	DepDelay	实际起飞时间与预计起飞时间之差
17	Origin	出发机场 (IATA 机场代码), 对应 airports.csv
18	Dest	到达机场 (IATA 机场代码), 对应 airports.csv

续表

变量编号	变量名	释义
19	Distance	出发机场与到达机场间距离 (单位: 英里)
20	TaxiIn	飞机起飞时滑行时间
21	TaxiOut	飞机降落时滑行时间
22	Cancelled	航班是否被取消
23	CancellationCode	航班被取消的原因 (A 为航空公司的原因, B 为天气原因, C 为国家航空系统的原因, D 为安全原因)
24	Diverted	飞机是否有改道, 1 为有改道
25	CarrierDelay	因航空公司原因导致的延误时长
26	WeatherDelay	因天气原因导致的延误时长
27	NASDelay	因国家航空系统原因导致的延误时长
28	SecurityDelay	因安全原因导致的延误时长
29	LateAircraftDelay	因晚飞导致的延误时长

关于数据变量说明, 最权威的信息来自如下网址:

http://www.transtats.bts.gov/printProfile.asp?DB_ID=120&Link=0

http://www.transtats.bts.gov/Fields.asp?Table_ID=236

需要说明的是, 原始数据是从 1987 年 10 月 14 日开始的, 本案例并不包含 1987 年的数据. 有兴趣的读者可以自行下载.

此外, 有两个辅助文件:airports.csv 包含了所有的机场信息, 字段说明见表 4.2; carriers.csv 给出了航空公司信息, 字段说明见表 4.3. 天气对航班延误有重大的影响, 因此作者编写爬虫程序从 http://www.wunderground.com/history 下载了相应时间、相应机场的天气信息, 存储在文件 rawweatherdata.csv 中, 字段说明见表 4.4. 读者可以在相应网站下载这些数据, 也可在中国人民大学出版社网站 (www.crup.com.cn) 下载, 本案例的所有程序均可在出版社网站下载.

表 4.2 airports.csv 字段说明

变量编号	变量名	释义
1	iata	国际机场缩写
2	airport	机场名称
3	city	机场所在城市
4	state	机场所在的州
5	country	机场所在国家 (少数几个不在美国境内)
6	lat	机场的纬度
7	long	机场的经度

表 4.3 carriers.csv 字段说明

变量编号	变量名	释义
1	Code	航空公司的代码
2	Description	航空公司的名称

表 4.4 天气数据字段说明

变量编号	变量名	变量类型	含义	单位
1	Yeartmp	离散变量	年份	—
2	Monthtmp	离散变量	月份	—
3	Daytmp	离散变量	日期	—
4	Maxtemp	连续变量	最高气温 (max temperature)	摄氏度
5	Meantemp	连续变量	平均气温 (mean temperature)	摄氏度
6	Mintemp	连续变量	最低气温 (min temperature)	摄氏度
7	Maxdewpoint	连续变量	最高露点① (max dew point)	摄氏度
8	Meandewpoint	连续变量	平均露点 (mean dew point)	摄氏度
9	Mindewpoint	连续变量	最低露点 (min dew point)	摄氏度
10	Maxhumidity	连续变量	最大湿度 (max humidity)	%
11	Meanhumidity	连续变量	平均湿度 (mean humidity)	%
12	Minhumidity	连续变量	最小湿度 (min humidity)	%
13	Maxsealevelpre	连续变量	最高海平面气压 (max sea level pressure)	百帕
14	Meansealevelpre	连续变量	平均海平面气压 (mean sea level pressure)	百帕
15	Minsealevelpre	连续变量	最低海平面气压 (min sea level pressure)	百帕
16	Maxvisibility	连续变量	最高能见度 (max visibility)	km
17	Meanvisibility	连续变量	平均能见度 (mean visibility)	km
18	Minvisibility	连续变量	最低能见度 (min visibility)	km
19	Maxwindspeed	连续变量	最大风速 (max wind speed)	km/h
20	Meanwindspeed	连续变量	平均风速 (mean wind speed)	km/h
21	Instantwindspeed	连续变量	瞬时风速 (instantaneous wind speed)	km/h
22	Rainfall	连续变量	降水量②	mm
23	Cloudcover	连续变量	云量③ (cloud cover)	—
24	Events	分类变量	活动④	—
25	Winddirdegrees	连续变量	风向⑤	度
26	Airporttmp	分类变量	机场	—
27	Cityabbr	分类变量	机场所在城市	—

注: ① 露点, 又称露点温度, 在气象学中是指在固定气压下, 空气中所含的气态水达到饱和而凝结成液态水所需降至的温度.
② 在原始网站上, 降水量中 T 表示微量降雨, 我们将其处理为 0.
③ 云量, 空中在视力范围内看到的云层的遮盖程度, 用 0～10 来表示.
④ 活动: 取值为 "中雨" "大雨" "大雾" 等.
⑤ 风向: 取值为 0～360 度的连续变量.

4.2 单机实现

4.2.1 基于 Mysql 的数据预处理

本案例使用的是数据库技术. 数据处理和分析的软件、语言、工具非常多, 各有优势和特点, 可以综合使用. 如果能够全部掌握, 当然非常理想. 受时间和精力所限, 很多时候只需熟练掌握其中一部分. 本书只介绍了一小部分目前主流的工具和方法. 随着时代的发展和技术的进步, 我们需要保持不断学习的能力.

Mysql 作为开源免费的数据库在处理结构化数据时具有很大的优势, 因此本案例的数据预处理采用 Mysql. 其他数据库技术与此类似.

步骤一: 利用 Shell 命令 (见下方程序 c0.sh) 将 1988—2008 年的 21 张表格合并为一张表格 (记为 airdata.csv). 此处我们将数据保存在 airdata 目录下, 读者可以根据自己的实际情况修改输入、输出文件的目录. 需要说明的是, 全部数据量较大, 运行速度较慢, 读者可以使用少数几个年份的数据进行调试.

```
#!/bin/bash
for ((i=1988;i<=2008;i++))
do
sed -i '1d' /airdata_temp/$i.csv #删除首行变量名
cat /airdata_temp/$i.csv /airdata_temp/airdata.csv >> /airdata_temp/airdata.csv
done
```

步骤二: 在 Mysql 中建立数据库和表格, 并将 21 年的航班数据 (即 airdata.csv) 导入 Mysql 中.

接下来利用 Mysql 和 Python 编写程序对数据进行描述统计分析. 此处程序 c1.sql 的目的是建表、拼接.

```
#####C1
create database airdata; #创建新的数据库
use airdata;
DROP TABLE IF EXISTS airdat;
CREATE TABLE airdat (
  `Year` int(255) DEFAULT NULL,
  `Month` tinyint(255) DEFAULT NULL,
  `DayofMonth` tinyint(255) DEFAULT NULL,
  `DayOfWeek` tinyint(255) DEFAULT NULL,
  `DepTime` varchar(255) DEFAULT NULL,
  `CRSDepTime` varchar(255) DEFAULT NULL,
  `ArrTime` varchar(255) DEFAULT NULL,
```

```sql
  `CRSArrTime` varchar(255) DEFAULT NULL,
  `UniqueCarrier` varchar(255) DEFAULT NULL,
  `FlightNum` varchar(255) DEFAULT NULL,
  `TailNum` varchar(255) DEFAULT NULL,
  `ActualElapsedTime` varchar(255) DEFAULT NULL,
  `CRSElapsedTime` varchar(255) DEFAULT NULL,
  `AirTime` varchar(255) DEFAULT NULL,
  `ArrDelay` varchar(255) DEFAULT NULL,
  `DepDelay` varchar(255) DEFAULT NULL,
  `Origin` varchar(255) DEFAULT NULL,
  `Dest` varchar(255) DEFAULT NULL,
  `Distance` varchar(255) DEFAULT NULL,
  `TaxiIn` varchar(255) DEFAULT NULL,
  `TaxiOut` varchar(255) DEFAULT NULL,
  `Cancelled` varchar(255) DEFAULT NULL,
  `CancellationCode` varchar(255) DEFAULT NULL,
  `Diverted` varchar(255) DEFAULT NULL,
  `CarrierDelay` varchar(255) DEFAULT NULL,
  `WeatherDelay` varchar(255) DEFAULT NULL,
  `NASDelay` varchar(255) DEFAULT NULL,
  `SecurityDelay` varchar(255) DEFAULT NULL,
  `LateAircraftDelay` varchar(255) DEFAULT NULL
) ENGINE=InnoDB DEFAULT CHARSET=ascii; #创建表格
LOAD DATA LOCAL INFILE '/home/air_data/airdata.csv'#文件路径(需加LOCAL以解决路径问题)
INTO TABLE airdat
CHARACTER SET ascii
FIELDS TERMINATED BY ',' ENCLOSED BY '"';#将数据导入Mysql
```

程序 c2.sql 对数据预处理.

```sql
####C2:
#各个年份的总航班数和延误数
select Year,count(*) as totalnum,count(if(ArrDelay>0,true,null)) as arrdelay from airdat
group by Year into outfile "/var/lib/mysql/total_delay1.txt";

#各个年份的平均出发延误和到达延误
select Year,avg(DepDelay) as depdelay, avg(ArrDelay) as arrdelay from airdat
group by Year into outfile "/var/lib/mysql/avg_delay1.txt";

注：outfile需注意是否有写入该路径的权限，一般默认的导出路径为/var/lib/mysql，可存储于该路径后复制
若希望修改存储目录权限，请参考：http://blog.csdn.net/lichangzai/article/details/1873328
```

利用如下程序 c2-2.py 画图.

```
###########################C2
#setwd("/home/air_data/")
library(reshape2)
library(ggplot2)

data1=read.table("total_delay1.txt",header=F)
colnames(data1)=c("year","总航班数","到达延迟航班数")
data1=melt(data1,id.vars="year")
ggplot(data1,aes(x=year,y=value))+geom_line(aes(x=year,colour=variable),size=2)+labs
(title="各个年份的总航班数和延迟航班数",x="年份",y="航班数")+ylim(0,8000000)

#各个年份的平均延迟时间
data2=read.table("avg_delay1.txt",header=F)
colnames(data2)=c("year","平均出发延迟","平均到达延迟")
data2=melt(data2,id.vars = "year")
ggplot(data2,aes(x=year,y=value))+geom_line(aes(x=year,colour=variable),size=2)+labs
(title="各个年份的平均延迟时间",x="年份",y="时间(min)")
```

首先我们关心的是各年份的总航班数以及到达延迟航班数 (见图 4.1). 接下来计算各年份航班的平均出发延误时间和平均到达延误时间 (见图 4.2).

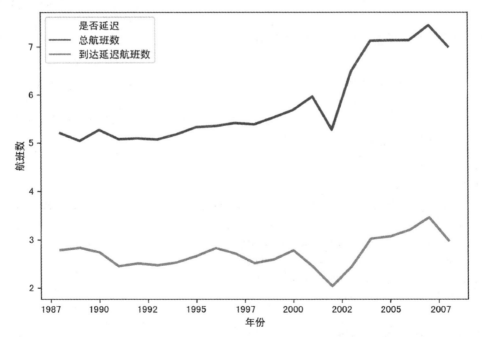

图 4.1 各个年份总航班数和到达延迟航班数

从图 4.1 中可以看出, 总航班数从 1988 年的 5 202 096 个上升到 2008 年的 7 009 728 个, 其中 2002 年有所减少, 航班延误率略小于 50%. 平均来讲, 航班出发延误在 5~11 分钟之间, 到达延误在 3~10 分钟之间, 略低于出发延误, 但低得非常有限, 只差 1 ~ 2 分钟.

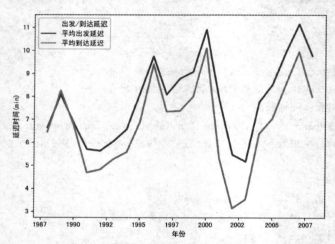

图 4.2　各个年份航班平均出发延迟时间和平均到达延迟时间

由于篇幅限制, 我们不给出更多的描述统计分析结果, 建议读者自行探索. 实际上, 对于任何数据分析, 尤其是大数据分析, 描述统计都是非常重要的, 它可以增进对数据的了解, 为准确的模型分析奠定基础.

4.2.2　洛杉矶到波士顿航线的延误分析

1. 数据预处理与描述统计

接下来以洛杉矶 (LAX) 到波士顿 (BOS) 的航线为例, 分析航班到达延误的原因. 程序 c3.sql 进行数据准备, 程序 c4.py 画图.

```
########C3
###洛杉矶到波士顿航线数据
CREATE TABLE laxtobos(select * from airdat where origin="LAX" and dest="BOS" and
Cancelled=0);
alter table laxtobos add column Delay varchar(20) not null after ArrDelay;
update laxtobos set Delay="1" where ArrDelay>0;
update laxtobos set Delay="0" where ArrDelay<=0;
select * from laxtobos limit 10;
```

```
#########################C4
install.packages("RODBC")
library(ggplot2)
library(RODBC)
```

```
library(plyr)
library(reshape2)
ch=odbcConnect("mydata","root","8888") #与airdata连接的数据库接口的名称和密码
#注:需下载对应数据库的odbc驱动,并在系统中进行设置,不同系统的具体设置方法可自行查阅

###各个月份的航班数量
data=sqlQuery(ch,"select Delay,Month from laxtobos")
names(data)
head(data)
data=prop.table(table(data),2)
write.csv(data,"E:/Machine Learning/lvxiaoling/process2/c2_各个月份航班数.txt")
x=barplot(data,main="各个月份的延误比例")
text(x,data[1,]-0.3,labels=round(data[1,],2))
text(x,data[1,]+0.07,labels=round(data[2,],2))

###每周各个工作日的航班数目和延误比例
data=sqlQuery(ch,"select Delay,DayOfWeek from laxtobos ")
data=prop.table(table(data),2)
write.csv(data,"E:/Machine Learning/lvxiaoling/process2/c2_每周各个工作日延误比例.txt")
x=barplot(data,main="每周各个工作日的延误比例")
text(x,data[1,]-0.3,labels=round(data[1,],2))
text(x,data[1,]+0.07,labels=round(data[2,],2))
```

首先从 airdata 表格中选取 1988—2008 年 LAX 到 BOS 所有未取消的航班数据, 并将其命名为 laxtobos. 此外, 根据 ArrDelay 是否大于 0, 将 ArrDelay 转化为延迟和不延迟的二分类变量 (ArrDelay $>$ 0 表示延迟, ArrDelay $<$ 0 表示不延迟).

汇总后, 可以看到 1988—2008 年间从 LAX 到 BOS 共有 54 346 个航班, 取消航班 32 个. 其中, 延误航班 (ArrDelay>0) 数目是 24 143 个, 占所有未取消航班数的 45.2%.

为了对数据有一个直观的认识, 同时也为了探索对航班延误有影响的因素, 我们先对 Year, Month, DayOfMonth, DayOfWeek, CRSDepTime, CRSArrTime, UniqueCarrier 几个变量进行描述统计分析. 部分结果展示如下:

(1) 各个月份的航班延误比率. 从图 4.3 中可以看出夏季 (6, 7, 8 月) 的航班延误率偏高, 因此可以初步认定 Month 对航班延误有影响.

(2) 每周各天的航班延误比率. 从图 4.4 中可以看出, 各个工作日的航班延误差别不大, 但周四、周五的航班延误率偏高, 周六偏低, 因此, 可以认为 DayOfWeek 对航班延误有影响.

此外, 我们对天气变量对航班延误的影响做了一些描述统计分析, 程序 c5.sql 的目的是加入天气变量生成数据, 进行预处理. 在此不给出具体分析结果.

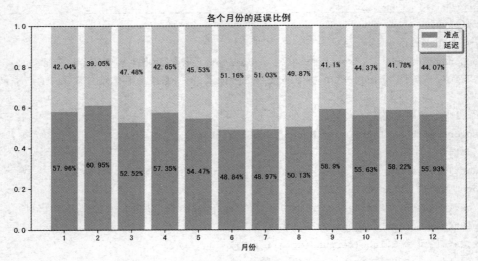

图 4.3　各个月份航班延误比例

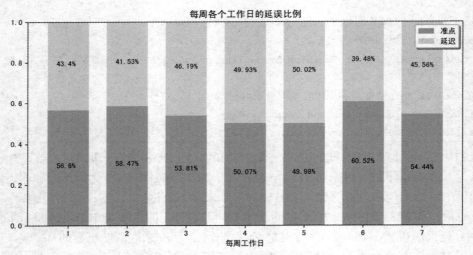

图 4.4　每周各天航班延误比例

```
########C5
##########导入天气变量
DROP TABLE IF EXISTS weatherdata;
CREATE TABLE weatherdata(
yeartmp INT NOT NULL,
monthtmp INT NOT NULL,
daytmp INT NOT NULL,
maxtemp FLOAT NULL,
meantemp FLOAT NULL,
mintemp FLOAT NULL,
```

```
maxdewpoint FLOAT NULL,
meandewpoint FLOAT NULL,
mindewpoint FLOAT NULL,
maxhumidity FLOAT NULL,
meanhumidity FLOAT NULL,
minhumidity FLOAT NULL,
maxsealevelpre FLOAT NULL,
meansealevelpre FLOAT NULL,
minsealevelpre FLOAT NULL,
maxvisibility FLOAT NULL,
meanvisibility FLOAT NULL,
minvisibility FLOAT NULL,
maxwindspeed FLOAT NULL,
meanwindspeed FLOAT NULL,
instantwindspeed FLOAT NULL,
rainfall FLOAT NULL,
cloudcover INT NULL,
events varchar(50) NULL,
winddirdegrees varchar(50) NULL,
airporttmp varchar(50) NOT NULL,
cityabbr varchar(50) NOT NULL)
ENGINE=InnoDB DEFAULT CHARSET=ascii; #创建表格

LOAD DATA INFILE 'F:/air/rawweatherdata.csv ' #文件路径
INTO TABLE weatherdata
CHARACTER SET ascii
FIELDS TERMINATED BY ','
(yeartmp, monthtmp, daytmp, @vmaxtemp, @vmeantemp, @vmintemp, @vmaxdewpoint,
@vmeandewpoint, @vmindewpoint, @vmaxhumidity, @vmeanhumidity, @vminhumidity,
@vmaxsealevelpre, @vmeansealevelpre, @vminsealevelpre, @vmaxvisibility,
@vmeanvisibility, @vminvisibility, @vmaxwindspeed, @vmeanwindspeed,
@vinstantwindspeed, @vrainfall, @vcloudcover, @vevents, @vwinddirdegrees, airporttmp,
cityabbr)
SET
maxtemp = nullif(@vmaxtemp,''),
meantemp = nullif(@vmeantemp,''),
mintemp = nullif(@vmintemp,''),
maxdewpoint=nullif( @vmaxdewpoint,'') ,
meandewpoint=nullif( @vmeandewpoint ,''),
mindewpoint=nullif( @vmindewpoint,''),
maxhumidity=nullif( @vmaxhumidity,''),
```

```
meanhumidity=nullif( @vmeanhumidity ,''),
minhumidity=nullif( @vminhumidity,''),
maxsealevelpre=nullif( @vmaxsealevelpre ,''),
meansealevelpre=nullif( @vmeansealevelpre,''),
minsealevelpre=nullif( @vminsealevelpre ,''),
maxvisibility=nullif( @vmaxvisibility,''),
meanvisibility=nullif( @vmeanvisibility,''),
minvisibility=nullif( @vminvisibility,''),
maxwindspeed=nullif( @vmaxwindspeed,''),
meanwindspeed=nullif( @vmeanwindspeed,''),
instantwindspeed=nullif( @vinstantwindspeed,''),
rainfall=nullif( @vrainfall,''),
cloudcover=nullif( @vcloudcover,''),
events =nullif( @vevents,''),
winddirdegrees=nullif( @vwinddirdegrees,'')

;
#创建表格laxtobos2,并将相关变量转化为分类变量
show tables;
create table laxtobos2 as (select Year,Month,DayofMonth,DayOfWeek,CRSDepTime,
CRSArrTime,UniqueCarrier,Origin,Dest,ArrDelay,DepDelay from laxtobos);
#########修改选取的数据结构
alter table laxtobos2 add column MonthFactor varchar(20) not null after Month;
update laxtobos2 set MonthFactor="Spring" where Month in (3,4,5);
update laxtobos2 set MonthFactor="Summer" where Month in (6,7,8);
update laxtobos2 set MonthFactor="Autumn" where Month in (9,10,11);
update laxtobos2 set MonthFactor="Winter" where Month in (12,1,2);

alter table laxtobos2 add column DayOfWeekFactor varchar(20) not null after DayOfWeek;
update laxtobos2 set DayOfWeekFactor="HeadWeek" where DayOfWeek in (1,2);
update laxtobos2 set DayOfWeekFactor="MediumWeek" where DayOfWeek in (3,4,5);
update laxtobos2 set DayOfWeekFactor="Saturday" where DayOfWeek in (6);
update laxtobos2 set DayOfWeekFactor="Sunday" where DayOfWeek in (7);

alter table laxtobos2 add column CRSDepTimeFactor varchar(20) not null after
CRSDepTime;
update laxtobos2 set CRSDepTimeFactor="Night" where CRSDepTime>=0000 and
CRSDepTime<0500;
update laxtobos2 set CRSDepTimeFactor="EarlyMorning" where CRSDepTime>=0500 and
CRSDepTime<0800;
```

```
update laxtobos2 set CRSDepTimeFactor="Morning" where CRSDepTime>=0800 and
CRSDepTime<1100;
update laxtobos2 set CRSDepTimeFactor="Noon" where CRSDepTime>=1100 and
CRSDepTime<1300;
update laxtobos2 set CRSDepTimeFactor="Afternoon" where CRSDepTime>=1300 and
CRSDepTime<1700;
update laxtobos2 set CRSDepTimeFactor="Evening" where CRSDepTime>=1700 and
CRSDepTime<2100;
update laxtobos2 set CRSDepTimeFactor="LateEvening" where CRSDepTime>=2100 and
CRSDepTime<2400;

alter table laxtobos2 add column CRSArrTimeFactor varchar(20) not null after
CRSArrTime;
update laxtobos2 set CRSArrTimeFactor="Night" where CRSArrTime>=0000 and
CRSArrTime<0500;
update laxtobos2 set CRSArrTimeFactor="EarlyMorning" where CRSArrTime>=0500 and
CRSArrTime<0800;
update laxtobos2 set CRSArrTimeFactor="Morning" where CRSArrTime>=0800 and
CRSArrTime<1100;
update laxtobos2 set CRSArrTimeFactor="Noon" where CRSArrTime>=1100 and
CRSArrTime<1300;
update laxtobos2 set CRSArrTimeFactor="Afternoon" where CRSArrTime>=1100 and
CRSArrTime<1700;
update laxtobos2 set CRSArrTimeFactor="Evening" where CRSArrTime>=1700 and
CRSArrTime<2100;
update laxtobos2 set CRSArrTimeFactor="LateEvening" where CRSArrTime>=2100 and
CRSArrTime<2400;

alter table laxtobos2 add column Delay varchar(20) not null after ArrDelay;
update laxtobos2 set Delay="1" where ArrDelay>0;
update laxtobos2 set Delay="0" where ArrDelay<=0;

####提取LAX的天气数据
create table laxweatherdata as
(select yeartmp as depyeartmp,monthtmp as depmonthtmp,daytmp as depdaytmp,
airporttmp as depairporttmp, meantemp as depmeantemp,meandewpoint as depmeandewpoint,
meanhumidity as depmeanhumidity, meansealevelpre as depmeansealevelpre,
meanvisibility as depmeanvisibility,meanwindspeed as depmeanwindspeed,
rainfall as deprainfall,cloudcover as depcloudcover
FROM weatherdata where airporttmp="LAX");
```

```
####提取BOS的天气数据
create table bosweatherdata as
(select yeartmp as arryeartmp,monthtmp as arrmonthtmp,daytmp as arrdaytmp,
airporttmp as arrairporttmp, meantemp as arrmeantemp,meandewpoint as arrmeandewpoint,
meanhumidity as arrmeanhumidity, meansealevelpre as arrmeansealevelpre,
meanvisibility as arrmeanvisibility,meanwindspeed as arrmeanwindspeed,
rainfall as arrrainfall,cloudcover as arrcloudcover
FROM weatherdata where airporttmp="BOS");

##处理异常值:云量的取值范围为0~10,超出此范围为缺失值
update bosweatherdata set arrcloudcover=NULL where arrcloudcover<0 or
arrcloudcover>10;

#将航空数据与出发机场数据合并
DROP TABLE IF EXISTS airweathertmp;
create table airweathertmp as (select a.*,b.* from laxtobos2 a LEFT JOIN
laxweatherdata b on
a.Year=b.depyeartmp AND a.Month=b.depmonthtmp AND a.DayOfMonth=b.depdaytmp);

#将航空数据与到达机场数据合并
DROP TABLE IF EXISTS airweather;
create table airweather as (select a.*,b.* from airweathertmp a LEFT JOIN
bosweatherdata b on
a.Year=b.arryeartmp AND a.Month=b.arrmonthtmp AND a.DayOfMonth=b.arrdaytmp);

#从合并的表格中提取我们分类模型用到的变量
DROP TABLE IF EXISTS mydata;
create table mydata as (select Delay, MonthFactor, DayOfWeekFactor, CRSDepTimeFactor,
CRSArrTimeFactor, UniqueCarrier, depmeantemp, depmeandewpoint, depmeanhumidity,
depmeansealevelpre, depmeanvisibility, depmeanwindspeed, deprainfall, depcloudcover,
arrmeantemp, arrmeandewpoint, arrmeanhumidity, arrmeansealevelpre, arrmeanvisibility,
arrmeanwindspeed, arrrainfall, arrcloudcover from airweather);
```

2. 建立分类模型预测航班延误

在以上洛极矶至波士顿未取消航班数据的基础上,选取天气变量中对应时间的出发机场和到达机场的 Meantemp, Meandewpoint, Meanhumidity, Meansealevelpre, Meanvisibility, Meanwindspeed, Rainfall, Cloudcover 几个变量作为自变量。出发机场的相应变量名加前缀 dep, 到达机场加 arr。此外,选取 Month, DayOfWeek, CRSDepTime, CRSArrTime 和 UniqueCarrier 作为自变量,将数据表命名为 laxtobos2, 对变量取值类别情况进行适当的汇总,具体转换规则如表 4.5 所示。

表 4.5 分类变量转换规则汇总

变量	变量名	释义
MonthFactor	Month	若 Month=12,1,2, 则 MonthFactor=winter
		若 Month=3,4,5, 则 MonthFactor=spring
		若 Month=6,7,8, 则 MonthFactor=summer
		若 Month=9,10,11, 则 MonthFactor=autumn
DayOfWeekFactor	DayOfWeek	若 DayOfWeek=1,2, 则 DayOfWeekFactor=headweek
		若 DayOfWeek=3,4,5, 则 DayOfWeekFactor=mediumweek
		若 DayOfWeek=6, 则 DayOfWeekFactor=saturday
		若 DayOfWeek=7, 则 DayOfWeekFactor=sunday
CRSDepTimeFactor	CRSDepTime	night(0:00—5:00),early morning(5:00—8:00),
		morning(8:00—11:00),noon(11:00—13:00),
		afternoon(13:00—17:00),evening(17:00—21:00),
		late evening(21:00—24:00)
CRSArrTimeFactor	CRSArrTime	同上
UniqueCarrierFactor	UniqueCarrier	仍然使用 UniqueCarrier 的分类

在剔除了 laxtobos2 数据的非完全样本后, 数据集样本数由 53 414 个降至 48 912 个. 我们选取 60% 的样本作为训练集, 其余作为测试集. 接下来, 用随机森林对上述数据进行分类. 在随机森林中, 构建每一棵决策树时, 随机选取 5 个变量, 一共构造 500 棵决策树. 用训练集构建随机森林, 各个变量的相对重要性见图 4.5 (该图通过下列程序 c5_2.py 画出).

```
##R代码
#install.packages("randomForest")
library(randomForest)
#install.packages("gbm")
library(gbm)
#install.packages("e1071")
library(e1071)
library(RODBC)
ch=odbcConnect("mydata","root","1000") #与airdata连接的数据库接口的名称和密码
#注:需下载对应数据库的odbc驱动,并在系统中进行设置,不同系统的具体设置方法可自行查阅

mydata=sqlQuery(ch,"select * from mydata")
tail(mydata)
mydata=mydata[complete.cases(mydata),]
dim(mydata)

mydata$Delay=as.factor(mydata$Delay)
mydata$MonthFactor=as.factor(mydata$MonthFactor)
```

```r
mydata$DayOfWeekFactor=as.factor(mydata$DayOfWeekFactor)
mydata$CRSDepTimeFactor=as.factor(mydata$CRSDepTimeFactor)
mydata$CRSArrTimeFactor=as.factor(mydata$CRSArrTimeFactor)
mydata$UniqueCarrier=as.factor(mydata$UniqueCarrier)

index2=function(table) {
  Accuracy=table[1,1]+table[2,2]
  precision=table[2,2]/sum(table[,2])
  recall=table[2,2]/sum(table[2,])
  F_measure=2*precision*recall/(precision+recall)#计算Recall, Precision和F-measure
  results=data.frame(Accuracy=Accuracy,recall=recall,precision=precision,F_measure=F_measure)
  return(results)
}
#将数据分为训练集和测试集
set.seed(100)
train=mydata[sample(1:dim(mydata)[1],ceiling(dim(mydata)[1]*0.6)),]
test=mydata[-sample(1:dim(mydata)[1],ceiling(dim(mydata)[1]*0.6)),]
#RF
summary(mydata$Delay)
rf.air=randomForest(Delay~.,data=train,mtry=5,importance=TRUE,ntree=500)
help(predict)
rf.pred=predict(rf.air,newdata=test[,-1]) ####有修改
rf.real=test$Delay
table_RF=table(rf.real,rf.pred)/nrow(test)
a=index2(table_RF)

#SVM_LINEAR
svmfit=svm(Delay~., data=train, kernel="linear", cost=1)
svm.pred=predict(svmfit,test)
svm.real=test$Delay
table(svm.real,svm.pred)
table_svm1=table(svm.real,svm.pred)/nrow(test)

#SVM_RADIAL
svmfit1=svm(Delay~., data=train, kernel="radial", cost=100)
svm.pred1=predict(svmfit1,test)
svm.real1=test$Delay
table(svm.real1,svm.pred1)
table_svm2=table(svm.real1,svm.pred1)/nrow(test)
```

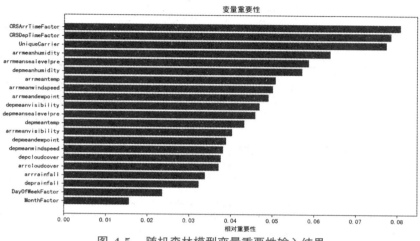

图 4.5 随机森林模型变量重要性输入结果

从图 4.5 中可以看出, 天气因素和时间因素 (月份、星期) 对航班延误都有较大影响. 利用训练后的模型对测试集进行预测, 得到混淆矩阵和各种评价指标 (见表 4.6 和表 4.7).

表 4.6 随机森林模型测试集结果 (1)

Real/Pred	0	1
0	1899(40.37%)	2802(14.32%)
1	3874(19.34%)	5080(25.96%)

表 4.7 随机森林模型测试集结果 (2)

	Accuracy	Recall	Precision	F_measure
RF	0.6634	0.5731	0.6445	0.6067

由表 4.7 可知, 从预测的整体准确率来看, 随机森林可以达到 66%, 效果还是比较好的.

我们使用随机森林作为基准的模型, 用每条样本上的预测概率进行排序后分组, 查看各组的预测效果, 可以看到, 选择靠前或靠后的组进行预测能够显著提升效果 (见表 4.8).

表 4.8 分组结果展示

组号	航班数	延迟数	延迟率	累积延迟率	预测准确率	累积预测准确率
1	1000	807	80.7%	80.7%	0.807	0.807
2	1000	725	72.5%	76.6%	0.725	0.766
3	1000	658	65.8%	73%	0.658	0.73
4	1000	633	63.3%	70.58%	0.633	0.7058
5	1000	555	55.5%	67.56%	0.555	0.6756
6	1000	507	50.7%	64.75%	0.514	0.6487
7	1000	465	46.5%	62.14%	0.535	0.6324
8	1000	421	42.1%	59.64%	0.579	0.6285
9	1000	393	39.3%	57.38%	0.607	0.6237
10	1000	363	36.3%	55.27%	0.637	0.625
11	1000	315	31.5%	53.1%	0.685	0.63

续表

组号	航班数	延迟数	延迟率	累积延迟率	预测准确率	累积预测准确率
12	1000	288	28.8%	51.08%	0.712	0.6372
13	1000	223	22.3%	48.87%	0.777	0.648
14	1000	187	18.7%	46.71%	0.813	0.6598
15	674	116	17.2%	45.35%	0.828	0.6675

接下来, 使用支持向量机对上述问题进行分类, 选取 radial 核函数, 并且取 cost 为 100, 结果见表 4.9 和表 4.10.

表 4.9 支持向量机模型测试集结果 (1)

Real/Pred	0	1
0	8996(44.56%)	2150(10.65%)
1	3370(16.69%)	5674(28.10%)

表 4.10 支持向量机模型测试集结果 (2)

	Accuracy	Recall	Precision	F_measure
SVM	0.7266	0.6274	0.7252	0.6728

我们还使用 Logistic 模型以及支持向量机 (线性核) 分析此数据, 测试集的预测准确性较低, 说明该问题中自变量对航班延误的影响是非线性的.

4.2.3 机场聚类分析

1. 数据处理与描述分析

下面利用 1988—2008 年的所有数据对机场进行聚类分析. 首先统计每个机场的起飞和到达航班总数 (N)、到达延迟超过 15 分钟的比例 (PDepDelay, 到达延迟超过 15 分钟的航班数除以该机场到达延迟非空值的航班数)、出发延迟超过 15 分钟的比例 (PArrDelay, 出发延迟超过 15 分钟的航班数除以该机场出发延迟非空值的航班数)、取消航班的比例 (PCancelled, 取消航班的数量除以该机场所有的航班数) 以及该机场所有航线平均距离 (AveDistance, 起飞机场或到达机场为该机场的 Distance 变量平均值). 变量含义如表 4.11 所示, 共有 338 个

表 4.11 聚类所用变量含义

变量名	变量含义	Min	Mean	Max
N	每个机场的起飞和到达航班总数	14	723214	13100431
PDepDelay	到达延迟超过 15 分钟的比例	1.50%	13.43%	55.56%
PArrDelay	出发延迟超过 15 分钟的比例	1.50%	19.13%	44.36%
PCancelled	取消航班的比例	0.00%	2.57%	20.06%
AveDistance	该机场所有航线平均距离	28	410.20	1707

机场. 这部分数据预处理的计算量较大, 使用单机版数据库技术比较耗时. 建议读者进行抽样处理, 这里不给出程序. 在 4.3 节中我们使用分布式处理方式给出这部分数据的预处理程序 (见程序 d2.hql). 在 Hive 下进行大数据处理比单机快得多.

对所有机场的各个变量进行描述统计, 通过频数直方图发现, 除出发延迟超过 15 分钟的比例 (PArrDelay) 这个变量, 其他各个变量均呈严重的右偏分布, 因此对其他变量分别取对数. 对于取消航班的比例 (PCancelled) 这个变量, 有三个机场取值为 0, 重新给这个变量赋值 —— 该变量除 0 以外的最小值, 然后再取对数. 为了消除聚类时量纲的影响, 对取完对数后的变量进行标准化处理. 取消航班的比例 (PCancelled) 的频数直方图如图 4.6 所示 (可通过下列程序 c6.R 来实现), 其他变量的图不再列出.

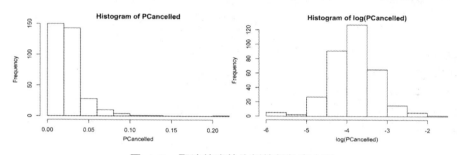

图 4.6　取消航班的比例的频数直方图

```
data<-read.csv("clusterdata.csv",sep="",col.names=c("origin","n_origin","n_dest",
"Cancel_origin",
"Cancel_dest","DepDelay","DepDelay_all","ArrDelay","ArrDelay_all","Distance_origin",
"nDistance_origin",
"Distance_dest","nDistance_dest"),header=F)
dim(data)###338个机场
data$iata<-data$origin
data$n<-data$n_origin+data$n_dest
data$PDepDelay<-data$DepDelay/data$DepDelay_all
data$PArrDelay<-data$ArrDelay/data$ArrDelay_all
data$PCancelled<-(data$Cancel_origin+data$Cancel_dest)/data$n
data$AveDistance<-(data$Distance_origin+data$Distance_dest)/(data$nDistance_dest+
data$nDistance_origin)
airport<-data[,(ncol(data)-5):ncol(data)]

sum(is.na(airport))
summary(airport)

a=airport
detach(a)
attach(a)
###右偏
```

```
mar = c(1, 2, 2, 2)
hist(n)
hist(PDepDelay)
hist(PArrDelay)
hist(PCancelled)
hist(AveDistance)
###原变量值为0的变量值赋值为该变量除0以外的最小值
which(a$PCancelled==0)
a$PCancelled[which(a$PCancelled==0)]<-a$PCancelled[which(a$PCancelled==0)]+
min(a[which(a$PCancelled!=0),]$PCancelled)
attach(a)
###取对数
hist(log(n))
hist(log(PDepDelay))
hist(log(PArrDelay))
hist(log(PCancelled))
hist(log(AveDistance))


a$n<-log(a$n)
a$PDepDelay<-log(a$PDepDelay)
a$PCancelled<-log(a$PCancelled)
a$AveDistance<-log(a$AveDistance)
detach(a)
attach(a)
head(a)
###kmeans
dist.a<-dist(scale(a[,2:6]))
plot(hclust(dist.a, method="ward"),labels=a[,1],xlab="airport",
     mar = c(13,2,6,1),main="Airport Cluster")
p<-NA
for(i in 2:6){
  model <- kmeans(scale(a[,2:6]),i,nstart =10)
  p[i-1]<-model$betweenss/model$totss
}
p
#轮廓系数
library(fpc)
silwidth<-NA
for (i in 2:6){
  model <- kmeans(scale(a[,2:6]),i,nstart=10)
  stats=cluster.stats(dist.a, model$cluster)
  silwidth[i-1]=stats$avg.silwidth#轮廓系数
```

```
}
silwidth
i=which.max(silwidth)+1;i
model <- kmeans(scale(a[,2:6]),3,nstart=10)
model
a$class<-model$cluster
model$size
model$betweenss/model$totss
a$sum<-NA
#插入机场地址数据
location<-read.csv("F:/air/airports.csv")
a<-merge(a,location, by = "iata",incomparables = NA)
M<-model$centers;M
#计算每个机场跟类中心的距离
a$sum[which(a$class==1)]<-rowSums((scale(a[,2:6])[which(a$class==1),]-M[1,])^2)
a$sum[which(a$class==2)]<-rowSums((scale(a[,2:6])[which(a$class==2),]-M[2,])^2)
a$sum[which(a$class==3)]<-rowSums((scale(a[,2:6])[which(a$class==3),]-M[3,])^2)
a1<-a[which(a$class==1),][which(rank(a[which(a$class==1),]$sum)<6),]
a2<-a[which(a$class==2),][which(rank(a[which(a$class==2),]$sum)<6),]
a3<-a[which(a$class==3),][which(rank(a[which(a$class==3),]$sum)<6),]
write.csv(a,file="class.csv")
###画图
library(maps)
#Airport Cluster of USA state
map('state',mar = c(3, 8, 6, 5), panel.first = grid(),col=9,cex=0.8)
axis(1, lwd = 0,cex=0.8);
axis(2, lwd = 0,cex=0.8);
title("Airport Cluster of USA state",cex=0.6)
points(long[which(a$class==1)],lat[which(a$class==1)], pch =22,col="blue",cex=0.8)
points(long[which(a$class==2)],lat[which(a$class==2)], pch =2,col="red",cex=0.8)
points(long[which(a$class==3)],lat[which(a$class==3)], pch =8,col="dark
green",cex=0.8)
attach(a)

#Airport Cluster of Alaska
map('world',c('USA','Alaska'),mar = c(3, 8, 6, 5),
    panel.first = grid(),ylim=c(50,75),xlim=c(-180,-120),col=9)
axis(1, lwd = 0);
axis(2, lwd = 0);
title("Airport Cluster of Alaska")
points(long[which(a$class==1)],lat[which(a$class==1)], pch =22,col="blue",cex=0.8)
points(long[which(a$class==2)],lat[which(a$class==2)], pch =2,col="red",cex=0.8)
```

```
points(long[which(a$class==3)],lat[which(a$class==3)], pch =8,col="dark
green",cex=0.8)

#Airport Cluster of Hawaii
map('world', c("USA","Hawaii"),mar = c(3, 8, 6, 5),
    ylim=c(18,23),xlim=c(-160,-153),col=9)
axis(1, lwd = 0);
axis(2, lwd = 0);
title("Airport Cluster of Hawaii")
points(long[which(a$class==1)],lat[which(a$class==1)], pch =22,col="blue",cex=0.8)
points(long[which(a$class==2)],lat[which(a$class==2)], pch =2,col="red",cex=0.8)
points(long[which(a$class==3)],lat[which(a$class==3)], pch =8,col="dark
green",cex=0.8)
legend("topright",c("class1","class2","class3"),pch=c(22,2,8),col=c("blue","red","dark
green"))
```

2. K 均值聚类

接下来使用 K 均值聚类算法对数据进行进一步的分析, 在程序中设定随机重复 10 次初始赋值. 此外, 在 K 均值方法中, 需要事先设定待分类的个数 K. 我们将 K 的取值依次设为 2, 3, 4, 5, 6, 分别根据组间方差占总方差的比重和轮廓系数来选择合理的 K 值. 轮廓系数 (Silhouette Coefficient) 可以用来判断聚类的优良性, 在 $-1 \sim +1$ 之间取值, 值越大表示聚类效果越好. 具体方法如下:

(1) 对于第 i 个元素 x_i, 计算 x_i 与同一个类内所有其他元素距离的平均值, 记作 a_i, 用于量化类内的凝聚度.

(2) 选取 x_i 外的一个类 b, 计算 x_i 与 b 中所有点的平均距离, 遍历所有其他类, 找到最近的这个平均距离, 记作 b_i, 用于量化类之间的分离度.

(3) 对于元素 x_i, 轮廓系数 $s_i = (b_i - a_i)/\max(a_i, b_i)$.

(4) 计算所有 x 的轮廓系数, 求出的平均值即当前聚类的整体轮廓系数.

从上面的公式可以看出, 若 s_i 小于 0, 说明 x_i 与类内元素的平均距离小于最近的其他类, 表示聚类效果不好; 如果 a_i 趋于 0, 或者 b_i 足够大, 那么 s_i 趋于 1, 说明聚类效果比较好.

在 R 语言中 fpc 包可以用于计算聚类后的一些评价指标, 其中包括轮廓系数. 采用 K 均值方法, 聚类结果如表 4.12 所示.

表 4.12 聚类分析结果

K 值	2	3	4	5	6
组间方差/总方差	0.244	0.439	0.506	0.562	0.601
轮廓系数	0.252	0.311	0.228	0.246	0.231

由表 4.12 可知, 当 K 值从 2 增加到 3 时, 组间方差的占比上升较快, 增加一个类, 组间方差/总方差可上升 19.5% 左右; 当 K 值大于 3 时, 组间方差/总方差的上升幅度较小, 约

5%; 当 $K=3$ 时, 轮廓系数最大. 因此, 综合考虑, 我们认为当 K 值取 3 时, 即将机场分为 3 类较为合适.

当 $K=3$ 时, 组间方差/总方差为 0.439, 各个类中机场数分别为 7 991 168 个. 相应的各类中心的柱状图如图 4.7 所示 (由 Excel 完成).

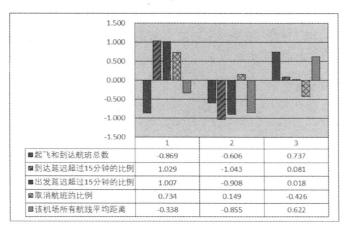

图 4.7　机场聚类类中心柱状图

类别 1 的机场, 其特点为航班总数最少, 所有航线平均距离较短, 而到达延迟比例、出发延迟比例、取消航班比例最多; 类别 2 的机场, 其特点为航班总数较少, 所有航线平均距离最短, 到达延迟比例、出发延迟比例最少, 取消航班比例较少; 类别 3 的机场, 其特点为航班总数最多, 所有航线平均距离最长, 到达延迟比例、出发延迟比例较少, 取消航班比例最少.

根据 2012 年美国机场统计报告, 2012 年美国大型枢纽机场旅客吞吐量排在前 29 名的机场中有 26 个机场在我们所分析的 338 个机场中, 并且这 26 个机场均被聚到类别 3 (可参考 http://www.krc.com.cn/ch/hygc/33/cid/82/id/258.html).

表 4.13 列出各类机场中离类中心最近的前五个机场.

表 4.13　各类机场中离类中心最近的前五个机场

class	iata	airport	city	state
1	CWA	Central Wisconsin	Mosinee	WI
	PIA	Greater Peoria Regional	Peoria	IL
	VLD	Valdosta Regional	Valdosta	GA
	SPI	Capital	Springfield	IL
	CEC	Jack McNamara	Crescent City	CA
2	ELM	Elmira/Corning Regional	Elmira	NY
	TWF	Joslin Field-Magic Valley	Twin Falls	ID
	UCA	Oneida County	Utica	NY
	CPR	Natrona County Intl	Casper	WY
	DLH	Duluth International	Duluth	MN
3	ABE	Lehigh Valley International	Allentown	PA
	MSN	Dane County Regional	Madison	WI
	HSV	Huntsville International	Huntsville	AL
	RIC	Richmond International	Richmond	VA
	LRD	Laredo International	Laredo	TX

4.2.4 最短路径

当两个城市之间没有直飞航班或可选的直飞航班很少时,通常的解决办法是选择转机.这一部分通过图的算法(具体介绍,见本书第 2 章),为没有直达航班的两个机场找到一条最短路径. 我们从图模型的角度来理解美国的航空规划问题,对应程序为 c7.R. 将美国所有的机场看做图上的顶点,将每段航线看做连接两个顶点的边,权重则可以根据要研究的具体问题定义为飞行时间、延误时间、距离等.

```
######2008年美国本土航线图
rm(list=ls())
library(maps)
library(igraph)
location <- read.csv('airports.csv', stringsAsFactors=F) #注意存储路径
test_time <- read.csv('2008.csv', stringsAsFactors = F) #注意存储路径


############ NA #############
ind <- which(is.na(test_time$Origin) | test_time$Dest == 0)
if (length(ind) !=0 ){
    test_time <- test_time[-ind,]
}
##########find longitude and latitude
test_time <- merge(test_time, location[, c('iata','state','lat', 'long')], by.x='Dest',
by.y='iata', all.x=T)
colnames(test_time)[(ncol(test_time)-2) : ncol(test_time)] <- c('Dest_ST','Dest_lat',
'Dest_long')

test_time <- merge(test_time, location[, c('iata','state', 'lat', 'long')], by.x='Origin',
by.y='iata', all.x=T)
colnames(test_time)[(ncol(test_time)-2) : ncol(test_time)] <- c('Origin_ST',
'Origin_lat', 'Origin_long')

##############construct graph data##########
airport_Origin <- unique(test_time [, c('Origin', 'Origin_lat', 'Origin_long')])
onlyDest <- setdiff(test_time$Dest, test_time$Origin)
if (length(onlyDest) != 0 ){
   allairport <- rbind(airport_Origin,
              unique(test_time[onlyDest, c('Origin', 'Origin_lat', 'Origin_long')]))
airport <- data.frame(unique (test_time[, c('Origin', 'Dest', 'Origin_lat',
                                  'Origin_long', 'Dest_lat', 'Dest_long')]))
}
```

```r
pdf('allPath_mainland.pdf', height=12, width=14) #注意存储路径
map('state')
for (j in 1:nrow(airport)){
   arrows(airport$Origin_long[j], airport$Origin_lat[j],
          airport$Dest_long[j], airport$Dest_lat[j],
          lwd=2,
          col='green')
}
for (i in 1:nrow(allairport)){
   points(x = allairport$Origin_long[i],
          y = allairport$Origin_lat[i])
   text(x = allairport$Origin_long[i],
        y = allairport$Origin_lat[i],
        allairport$Origin[i],
        col='blue',pos=2)
}
dev.off()

###########2008年美国境内航线图

pdf('allPath_all.pdf', height=12, width=14) #注意存储路径
m <- map("state", interior = FALSE, plot=FALSE)
###change names to abbs
data(state.fips)
m$names <- as.vector(state.fips$abb)
m.world <- map("world", c("USA", 'hawaii'), xlim=c(-180,-60), ylim=c(15,72), interior = FALSE)
map("state", add=T)
map("world", c("USA:Alaska"), add=T)
for (j in 1:nrow(airport)){
   arrows(airport$Origin_long[j], airport$Origin_lat[j],
          airport$Dest_long[j], airport$Dest_lat[j],
          lwd=2,
          col='green')
}
for (i in 1:nrow(allairport)){
   points(x = allairport$Origin_long[i],
          y = allairport$Origin_lat[i])
   text(x = allairport$Origin_long[i],
        y = allairport$Origin_lat[i],
        allairport$Origin[i],
        col='blue',pos=2)
```

```r
}
dev.off()

###############2008年飞往阿拉斯加航线
############### to Alaska ##############
AKrange <- which(test_time$Dest_ST == 'AK'& test_time$Origin_ST != 'AK')
test_time <- test_time[AKrange, ]

##############construct graph data##########
airport_Origin <- unique(test_time [, c('Origin', 'Origin_lat', 'Origin_long')])
onlyDest <- setdiff(test_time$Dest, test_time$Origin)
if (length(onlyDest) != 0 ){
   allairport <- rbind(airport_Origin,
                 unique(test_time[onlyDest, c('Origin', 'Origin_lat', 'Origin_long')]))
}

airport <- data.frame(unique (test_time[, c('Origin', 'Dest', 'Origin_lat',
                                            'Origin_long', 'Dest_lat', 'Dest_long')]))

toAK <- unique(test_time$Origin_ST)

pdf('allPath_toAL.pdf', height=12, width=14) #注意存储路径
###change names to abbs
data(state.fips)
toAK <- state.fips[which(state.fips$abb %in% toAK), 'polyname']
map("world", c("USA", 'hawaii', 'Alaska'), xlim=c(-180,-65), ylim=c(17,72), interior = FALSE)
map("state", add=T, col = 'grey', fill = T)
map('state', regions = toAK, col = 'white', fill=T, add=T)
for (j in 1:nrow(airport)){
   arrows(airport$Origin_long[j], airport$Origin_lat[j],
          airport$Dest_long[j], airport$Dest_lat[j],
          lwd=2,
          col='green')
}
for (i in 1:nrow(allairport)){
   points(x = allairport$Origin_long[i],
          y = allairport$Origin_lat[i])
   text(x = allairport$Origin_long[i],
        y = allairport$Origin_lat[i],
        allairport$Origin[i],
        col='blue',pos=2)
```

```r
}
dev.off()

####最短路径

rm(list=ls())
library(igraph)
library(maps)
set.seed(1234)

##############sampling
years <- paste(1988:2008, 'csv', sep='.')
JFKtoANC <- list()
for (i in 1:length(years)) {
    data <- read.csv(years[i], stringsAsFactors=F) #注意存储路径
    valid_data_date <- which(data$Month == 7 & data$DayofMonth == 4 |
                             data$Month == 7 & data$DayofMonth == 5)
    data <- data[valid_data_date, ]
    valid_data_path <- which(data$Origin == 'JFK'| data$Dest == 'ANC')
    JFKtoANC [[i]] <- data[valid_data_path, ]
    rm(data)
}
test_time <- do.call(rbind.data.frame, JFKtoANC)

#########NA data##########
ind <- which(is.na(test_time$CRSDepTime) | is.na(test_time$ArrTime))
test_time <- test_time[-ind,]

######arr >24 to the second day
ind.mis <- which(test_time$ArrTime < test_time$CRSDepTime)
test_time$DayofMonth_Arr <- test_time$DayofMonth
test_time$DayOfWeek_Arr <- test_time$DayOfWeek
test_time$DayofMonth_Arr[ind.mis] <- test_time$DayofMonth_Arr[ind.mis]+1
test_time$DayOfWeek_Arr[ind.mis] <- test_time$DayOfWeek[ind.mis]+1

##########find longitude and latitude
location <- read.csv('airports.csv', stringsAsFactors=F) #注意存储路径

test_time <- merge(test_time, location[, c('iata', 'lat', 'long')], by.x='Dest',
by.y='iata', all.x=T)
colnames(test_time)[(ncol(test_time)-1) : ncol(test_time)] <- c('Dest_lat', 'Dest_long')
```

```
test_time <- merge(test_time, location[, c('iata', 'lat', 'long')], by.x='Origin',
by.y='iata', all.x=T)
colnames(test_time)[(ncol(test_time)-1) : ncol(test_time)] <- c('Origin_lat',
'Origin_long')


#######sampling########transfer=1#########
#####Origins on 5th, with CRSDepTime =24 but ArrTime<24 should be omitted
ind <- which(test_time$DayofMonth == 5
             & test_time$ArrTime < test_time$CRSDepTime)
test_time <- test_time[-ind, ]
whole <- test_time

duration <- c(1988 : 2008)
year <- 1
cascade <- list()

for (c in duration) {
  test_time <- whole[which(whole$Year == c), ]
  unique.transfer <- unique(test_time$Dest)
  omit.ANC <- which(unique.transfer == 'ANC')
  unique.transfer <- unique.transfer[-omit.ANC]
  cascade.year <- list()
  for (i in 1: length(unique.transfer)) {
    ind.unique <- which(test_time$Dest == unique.transfer[i])
    for (j in 1:length(ind.unique)) {
      transfer <- ind.unique[j]
      ind <- which(test_time$Origin == test_time$Origin[transfer])
      if (length(ind)!=0) {
         if (test_time$DayofMonth_Arr[transfer] == 5) {
            cascade.1 <- which(test_time$Origin == test_time$Dest[transfer]
                               & test_time$DayofMonth == 5
                               & test_time$CRSDepTime >=
test_time$CRSArrTime[transfer])
         } else {
            cascade.1 <- which(test_time$Origin == test_time$Dest[transfer]
                        & ((test_time$DayofMonth == 4
                            & test_time$CRSDepTime >=
test_time$CRSArrTime[transfer])
                           | test_time$DayofMonth == 5))
         }
         if (length(cascade.1) != 0) {
            cascade.year[[i]] <- test_time[c(transfer, cascade.1), ]
```

```
                break
            }
        }
    }
    cascade[[year]] <- do.call(rbind.data.frame, cascade.year)
    year = year+1
}

test_time <- do.call(rbind.data.frame, cascade)
test_time$DepTime <- formatC(test_time$DepTime, width=4, flag=0)
test_time$ArrTime <- formatC(test_time$ArrTime, width=4, flag=0)
test_time$CRSDepTime <- formatC(test_time$CRSDepTime, width=4, flag=0)
test_time$CRSArrTime <- formatC(test_time$CRSArrTime, width=4, flag=0)
test_time <- transform(test_time, Arr.Date=as.Date(paste(Year, Month, DayofMonth_Arr,
    sep='-')))
test_time <- transform(test_time, CRSDep.Date=as.Date(paste(Year, Month, DayofMonth,
    sep='-')))

attach(test_time)
##################attach weight #############
weight <- array(0, nrow(test_time))
weight_comb <- matrix('', nrow(test_time), 6)

ind.JFK <- which(Origin == 'JFK')
for (c in duration) {
  ind <- which(Year == c)
  c.JFK <- ind.JFK[which(ind.JFK %in% ind)]
  for (i in c.JFK){
    weight[i] <- CRSElapsedTime[i]
    weight_comb[i,] <- c(Origin[i], Dest[i], Origin_lat[i], Origin_long[i],
                         Dest_lat[i], Dest_long[i])
  }
  no.JFK <- setdiff(ind, c.JFK)
  no.JFK.list <- unique( Origin [no.JFK])
  for (i in 1:length(no.JFK.list)) {
    JFK.arr <- which(Dest == no.JFK.list[i])
    JFK.arr <- ind[which(ind %in% JFK.arr)]
    JFK.leave <- which( Origin == no.JFK.list[i])
    JFK.leave <- ind[which(ind %in% JFK.leave)]
    for (j in JFK.leave){
      weight[j] <- CRSElapsedTime[j]
```

```r
         weight[j] <- weight[j] + difftime(strptime(paste(CRSDep.Date[j],
CRSDepTime[j]), '%Y-%m-%d %H%M'),
                                        strptime(paste(Arr.Date[JFK.arr],
ArrTime[JFK.arr]), '%Y-%m-%d %H%M'), units='mins')
         weight_comb[j,] <- c(Origin[j], Dest[j], Origin_lat[j], Origin_long[j],
                              Dest_lat[j], Dest_long[j])
      }
   }
}
colnames(weight_comb) <- c('Origin', 'Dest', 'Origin_lat', 'Origin_long',
                           'Dest_lat', 'Dest_long')
weight_comb <- data.frame(Origin = weight_comb[, 'Origin'],
                          Origin_lat = as.numeric(weight_comb[, 'Origin_lat']),
                          Origin_long = as.numeric(weight_comb[, 'Origin_long']),
                          Dest = weight_comb[, 'Dest'],
                          Dest_lat = as.numeric(weight_comb[, 'Dest_lat']),
                          Dest_long = as.numeric(weight_comb[, 'Dest_long']))

###average weight
unique_combination <- unique(weight_comb)
comb_weight <- matrix(, nrow(unique_combination))
for (i in 1:nrow(unique_combination)){
   ind <- which(weight_comb[,'Origin']== unique_combination[i, 1] &
                  weight_comb[,'Dest'] == unique_combination[i, 2] )
   comb_weight[i] <- sum(weight[ind])/length(ind)
}
whole <- test_time
test_time <- unique_combination

#################plot
library(maps)
library(igraph)
airport <- data.frame(airport = unique (unlist (test_time[, c('Origin', 'Dest')] )))
delay <- data.frame(from = test_time[, 'Origin'],
                    to = test_time[, 'Dest'],
                    Origin.lat = test_time[, 'Origin_lat'],
                    Origin.long = test_time[, 'Origin_long'],
                    Dest.lat = test_time[, 'Dest_lat'],
                    Dest.long = test_time[, 'Dest_long'])
g <- graph_from_data_frame(delay, directed=T)
E(g)$weight <- comb_weight
shortest <- shortest_paths(g, from='JFK', to='ANC')$vpath
```

```r
optimal.transfer <- as.vector(shortest[[1]][2])
all.path <- all_simple_paths(g, which(V(g)$name=='JFK'), which(V(g)$name=='ANC'))

pdf('allPath_minTotal.pdf', height=12, width=14) #注意存储路径
iArrows <- igraph:::igraph.Arrows
m <- map("state", interior = FALSE, plot=FALSE)
###change names to abbs
data(state.fips)
m$names <- as.vector(state.fips$abb)
m.world <- map("world", c("USA","hawaii"), xlim=c(-180,-65), ylim=c(19,72),interior = FALSE)
map("state", boundary = TRUE, add = TRUE, fill=F, interior=T )
#map("world", c("hawaii"), boundary = TRUE, add = TRUE)
map("world", c("USA:Alaska"), boundary = TRUE, add = TRUE)
for (i in 1:length(all.path)) {
  path.airport <- V(g)$name[as.vector(all.path[[i]])]
  airport1 <- which(test_time[, 'Origin'] == path.airport[1])[1]
  airport2 <- which(test_time[, 'Origin'] == path.airport[2])[1]
  airport3 <- which(test_time[, 'Dest'] == path.airport[3])[1]
  points(x=test_time[airport1, 'Origin_long'],
         y=test_time[airport1, 'Origin_lat'],
         col='red',cex=2)
  points(x=test_time[airport2, 'Origin_long'],
         y=test_time[airport2, 'Origin_lat'],
         col='red',cex=2)
  text(x=test_time[airport2, 'Origin_long'],
       y=test_time[airport2, 'Origin_lat'],
       path.airport[2],
       col='blue',pos=2)
  if (path.airport[2] == V(g)$name[optimal.transfer]){
     iArrows(test_time[airport1, 'Origin_long'], test_time[airport1, 'Origin_lat'],
             test_time[airport2, 'Origin_long'], test_time[airport2, 'Origin_lat'],
             h.lwd=2, sh.lwd=5, sh.col='green',
             curve=1 - (i %% 2), sh.lty=2, width=1, size=0.7)
     points(x=test_time[airport3, 'Dest_long'],
            y=test_time[airport3, 'Dest_lat'],
            col='red',cex=2)
     iArrows(test_time[airport2, 'Origin_long'], test_time[airport2, 'Origin_lat'],
             test_time[airport3, 'Dest_long'], test_time[airport3, 'Dest_lat'],
             h.lwd=2, sh.lwd=5, sh.col='green',
             curve=1 - (i %% 2), sh.lty=2, width=1, size=0.7)
```

```
    } else{
       arrows(test_time[airport1, 'Origin_long'], test_time[airport1, 'Origin_lat'],
              test_time[airport2, 'Origin_long'], test_time[airport2, 'Origin_lat'],
              lwd=2)
       points(x=test_time[airport3, 'Dest_long'],
              y=test_time[airport3, 'Dest_lat'],
              col='red',cex=2)
       Arrows(test_time[airport2, 'Origin_long'], test_time[airport2, 'Origin_lat'],
              test_time[airport3, 'Dest_long'], test_time[airport3, 'Dest_lat'],
              h.lwd=2, sh.lwd=2,
              curve=0.5 - 1.2*(i %% 2), width=1, size=0.7)
    }
}
text(x=test_time[airport1, 'Origin_long'],
     y=test_time[airport1, 'Origin_lat'],
     path.airport[1],
     col='blue', pos=2)
dev.off()
```

1. 描述统计

以 2008 年数据为例, 看一下美国全年的航班情况. 当年共 303 个机场有航班起飞记录, 304 个机场有航班降落记录. 其中, 科罗拉多州的 PUB 机场仅有两趟航班起飞记录, 怀俄明州的 CYS 机场和犹他州的 OGD 机场各有两趟航班降落记录. 在超过 90 000 个可能的航段里, 仅有 5 633 个航段有飞行记录. 其中, 旧金山 SFO 机场飞往洛杉矶 LAX 机场的航班数最多, 全年共有 13 788 个; 另有 285 个航段全年仅有一次飞行记录, 如虽然纽约 JFK 机场 2008 年有逾 20 万架飞机起飞或降落, 南卡罗来纳州 CHS 机场全年有近 3 万条飞机起飞或降落记录, 但 2008 年从 JFK 机场飞往 CHS 机场的航班只有一个.

从 2008 年美国本土航线图上看, 东部地区以及西部沿海地区明显机场更多, 航线相对更密集; 北部的航空明显没那么繁忙. 注意指向本土外东南、西南和西北方向的箭头, 这三个地方分别对应美属维尔京群岛、夏威夷和阿拉斯加.

美国其他州飞往阿拉斯加的航线并不多. 事实上, 2008 年全美本土仅有 14 个州 (包括夏威夷) 开通了 17 条直飞阿拉斯加的航线, 这 17 条航线全年共有 15 064 条飞行记录, 目的地主要是阿拉斯加的 ANC 机场.

2008 年美国其他州飞往阿拉斯加的全部航线共 17 条. 这 17 个始发的城市分别为亚特兰大、卡温顿、丹佛、达拉斯、底特律、火奴鲁鲁、休斯敦、拉斯维加斯、洛杉矶、明尼阿波利斯、卡胡卢伊、芝加哥、波特兰、凤凰城、西雅图、旧金山和盐湖城. 东部沿海地区、中部地区以及北部地区几乎没有航班直飞阿拉斯加.

2. 最短路径选择

现假定某旅客计划在当地时间 7 月 4 日从纽约 JFK 机场出发, 希望最晚在当地时间 7 月 5 日到达阿拉斯加 ANC 机场, 飞行途中最多允许转机一次. 他希望选择一条总时间 (飞行时间加上在机场等候的时间) 最短的路径.

我们首先在全部航班数据中找到每年 7 月 4 日或 5 日从纽约 JFK 机场出发的航班, 以及到达 ANC 机场的航班的记录.

在建立模型之前首先要保证样本中的飞行计划是可行的. 具体来说, 需要确保转机后所乘坐的航班的预计起飞时间在前一趟航班的预计到达时间之后. 根据这一原则选取符合要求的样本单元.

取得样本后, 给每一条航线设定权重. 由于旅客希望花费的总时间最短, 因此我们将第一段航线 (从纽约 JFK 机场到第一个转机点) 的权重定为这一航线上飞机的平均飞行时间, 另一段航线的权重定为这一航线上飞机的飞行时间和转机等候时间之和的平均.

将全部机场看做顶点, 机场间的航线为连接两个顶点的边, 可以构造一个图, 权重是两个机场之间各个航线的权重的平均. 通过 Dijkstra 算法求得纽约 JFK 机场到阿拉斯加 ANC 机场所花时间最短的路径.

在本例中, 共有 16 个机场可供转机, 分别为亚特兰大 ATL 机场、辛辛那提 CVG 机场、丹佛 DEN 机场、达拉斯 – 沃斯堡 DFW 机场、底特律大都会 DTW 机场、乔治布什 IAH 机场、拉斯维加斯麦卡伦 LAS 机场、洛杉矶 LAX 机场、明尼阿波利斯 – 圣保罗 MSP 机场、芝加哥奥黑尔 ORD 机场、波特兰 PDX 机场、凤凰天港 PHX 机场、西雅图 – 塔科马 SEA 机场、旧金山 SFO 机场、盐湖城 SLC 机场和兰伯特 – 圣路易斯 STL 机场. 而在辛辛那提 CVG 机场转机的路径为总花费时间最短的路径.

回到数据来看, 从纽约 JFK 机场至辛辛那提 CVG 机场这一航线从 1988 年开始每天有至少 1 个航班, 随着时间的推移航班数增加, 至 2008 年每天有 5 个航班, 一般为早晚各 1 班、下午 3 班, 下午的航班一般在 4 点左右到达 CVG 机场. 从辛辛那提 CVG 机场至阿拉斯加 ANC 机场的航线则是一条相对 "年轻" 的航线, 从 2006 年开始才有运营记录, 一般为每天一班, 起飞时间为下午 5 点左右. 选择这一联程航线, 旅客在辛辛那提 CVG 机场转机等候的时间较短, 让这个联程航线成为所有联程航线中总耗时最短的一个.

这里展示的只是应用图的方法寻找最短路径的一个简单的例子. 个人可以根据自己的喜好、价格等因素作出选择, 而不需要这么精准的数据分析. 但是, 身处大数据时代, 我们有全部航班的历史记录, 从航班管理者的角度完全可以应用这个算法评估各种转机方案的优劣, 设计更好的航线.

4.3 分布式实现

4.3.1 基于 Hive 的数据预处理

(1) 基于 2000—2008 年数据的起飞延误的分类模型数据预处理。因为数据量很大，用 Hive 来替代之前的 Mysql。虽然两者的语法十分相近，但是 Hive 的效率显著高于 Mysql。对应程序为如下所示的 d1.hql。所选用的变量有 month, dayofweek, crsdeptime, crsarrtime, distance, depmeantemp, depmeandewpoint, depmeanhumidity, depmeansealevelpre, depmeanvisibility, depmeanwindspeed, deprainfall, depcloudcover。筛选出 2000—2008 年的样本量约 5930 万条，其中完整数据约有 4850 万条。接下来利用完整数据进行模型分析。

```
程序d1: 起飞延误分类模型数据预处理
#################################Part1#################################
#在airdata数据库下建立表格airdat,并将航空数据导入表格中
hive #进入hive
CREATE database airdata;
use airdata;
CREATE TABLE airdat(year INT,month INT,dayofmonth INT,dayofweek INT,deptime
INT,crsdeptime INT,arrtime INT,crsarrtime INT,uniquecarrier STRING,flightnum
STRING,tailnum STRING,actualelapsedtime INT,crselapsedtime INT, airtime INT,arrdelay
INT ,depdelay INT,origin STRING,dest STRING,distance INT,taxiin INT,taxiout INT
,cancelled INT,cancellationcode STRING,diverted INT,carrierdelay INT ,weatherdelay
INT,nasdelay INT,securitydelay INT,lateaircraftdelay INT)
COMMENT "this is all of the 21 sheets"
ROW FORMAT DELIMITED FIELDS TERMINATED BY ','
STORED AS TEXTFILE;

load data local inpath '/home/air_data/airdata.csv' into table airdat;

###在hive中创建weatherdata表格，并将天气数据导入表格，作为出发机场的天气数据
CREATE TABLE weatherdata(yeartmp SMALLINT,monthtmp TINYINT,daytmp TINYINT,maxtemp
FLOAT,meantemp FLOAT,mintemp FLOAT,maxdewpoint
FLOAT,meandewpoint FLOAT,mindewpoint FLOAT ,maxhumidity FLOAT,meanhumidity FLOAT
,minhumidity FLOAT ,maxsealevelpre FLOAT,meansealevelpre FLOAT,minsealevelpre
FLOAT,maxvisibility FLOAT,meanvisibility FLOAT,minvisibility FLOAT,maxwindspeed
FLOAT,meanwindspeed FLOAT,instantwindspeed FLOAT,rainfall FLOAT,cloudcover INT,events
STRING,winddirdegrees SMALLINT,airporttmp STRING,cityabbr STRING)
```

```
COMMENT "this is all weather data of all airports during 1988-2008"
ROW FORMAT DELIMITED FIELDS TERMINATED BY ','
STORED AS TEXTFILE;

load data local inpath "/home/air_data/rawweatherdata.csv" into table weatherdata;

#############################Part2#################################
###从航空数据(airdat)中选取出建模需要的变量,并将其放在air表格中,用于之后的拼接表格
CREATE VIEW air
AS SELECT year,month,dayofmonth,dayofweek,crsdeptime,crsarrtime,uniquecarrier,origin,
dest,distance,arrdelay,depdelay
FROM airdat;

###从weatheradta中选取出发机场天气数据建模所需变量,并将其放在depweather表格中,用于之
后的拼接表格
CREATE VIEW depweather
AS SELECT yeartmp as depyeartmp,monthtmp as depmonthtmp,daytmp as depdaytmp,airporttmp
as depairporttmp,meantemp as depmeantemp,meandewpoint as depmeandewpoint,meanhumidity
as depmeanhumidity,meansealevelpre as depmeansealevelpre,meanvisibility as
depmeanvisibility,meanwindspeed as depmeanwindspeed,rainfall as deprainfall,cloudcover
as depcloudcover
FROM weatherdata;

###从到达机场天气数据(arrweatherdata)中选取建模所需变量,并将其放在arrweather表格中,
用于之后的拼接表格
CREATE VIEW arrweather
AS SELECT yeartmp as arryeartmp,monthtmp as arrmonthtmp,daytmp as arrdaytmp,airporttmp
as arrairporttmp,meantemp as arrmeantemp,meandewpoint as arrmeandewpoint,meanhumidity
as arrmeanhumidity,meansealevelpre as arrmeansealevelpre,meanvisibility as
arrmeanvisibility,meanwindspeed as arrmeanwindspeed,rainfall as arrrainfall,cloudcover
as arrcloudcover
FROM weatherdata;

##################Part3:合并############################
###将航空数据(air)与出发机场天气数据(depweather)进行合并,合并为tmp1
CREATE VIEW tmp1
AS SELECT a.*,b.*
FROM air a LEFT OUTER JOIN depweather b on a.origin=b.depairporttmp AND
a.year=b.depyeartmp AND a.month=b.depmonthtmp AND a.dayofmonth=b.depdaytmp;
```

```
###将tmp1与到达机场天气数据(arrweather)进行合并,合并为tmp2
CREATE VIEW tmp2
AS SELECT a.*,b.*
FROM tmp1 a LEFT OUTER JOIN arrweather b on a.dest=b.arrairporttmp AND
a.year=b.arryeartmp AND a.month=b.arrmonthtmp AND a.dayofmonth=b.arrdaytmp;

###退出hive环境,在linux下从合并好的表格中提取建立分类模型所需要的相关变量,将数据输出
到当下目录的classdata0008.csv。
exit;

hive -S -e"SELECT arrdelay, month, dayofweek, crsdeptime, crsarrtime, distance,
depmeantemp, depmeandewpoint, depmeanhumidity,
depmeansealevelpre, depmeanvisibility, depmeanwindspeed, deprainfall, depcloudcover,
arrmeantemp, arrmeandewpoint,
arrmeanhumidity, arrmeansealevelpre, arrmeanvisibility, arrmeanwindspeed, arrrainfall,
arrcloudcover FROM airdata.tmp2 where year > 1999" > classdata0008.csv
```

(2) 机场聚类数据预处理. 对机场聚类部分的数据使用 Hive 进行预处理. 对应程序为如下所示的 d2.hql. 因为数据处理后只有 338 行 13 列, 所以在接下来的分布式分析中不再给出程序. 下面仅给出分布式实现分类模型的示例.

```
程序d2: 机场聚类数据预处理
hive #进入hive
use airdata;
show tables;
###查询时显示列名
set hive.cli.print.header=true;
###设置每行显示列数
set hive.sample.seednumber=0;
###考察数据有多少行
select count(*) from airdata.airdata;
###观察前2行数据
select * from airdata.airdata limit 2;

create view airdat_new as select ArrDelay,DepDelay,Origin,Dest,Distance,Cancelled from
airdata.airdat;

###进行变量的提取
### 每个机场出发数量和到达数量
create view t1 as select origin,count(*) as n_origin from airdata.airdat_new group by
origin; create view t2 as select dest,count(*) as n_dest from airdata.airdat_new group
by dest;
```

```
###作为出发机场的取消数量和作为到达机场的取消数量
create view t3 as select origin,sum(Cancelled) as Cancel_origin from
airdata.airdat_new where Cancelled=1 or Cancelled=0 group by origin;
create view t4 as select dest,sum(Cancelled) as Cancel_dest from airdata.airdat_new
where Cancelled=1 or Cancelled=0 group by dest;
##作为出发机场时，延迟数量以及延迟超过15分钟数量
create view t5 as select origin,count(*) as DepDelay from airdata.airdat_new where
DepDelay>15 group by origin;
create view t6 as select origin,count(*) as DepDelay_all from airdata.airdat_new where
DepDelay is not null group by origin;
##作为到达机场时，延迟数量以及延迟超过15分钟数量
create view t7 as select dest,count(*) as ArrDelay from airdata.airdat_new where
ArrDelay>15 group by dest;
create view t8 as select dest,count(*) as ArrDelay_all from airdata.airdat_new where
ArrDelay is not null group by dest;
##作为出发机场和到达机场时航线总里程和具有航线里程的航班数量（用于计算平均值）
create view t9 as select origin,sum(Distance) as Distance_origin from
airdata.airdat_new group by origin;
create view t10 as select origin,count(*) as nDistance_origin from airdata.airdat_new
where Distance is not null group by origin;
create view t11 as select dest,sum(Distance) as Distance_dest from airdata.airdat_new
group by dest;
create view t12 as select dest,count(*) as nDistance_dest from airdata.airdat_new
where Distance is not null group by dest;
create view clusterdata as Select t1.*,t2.n_dest,t3.Cancel_origin,t4.Cancel_dest,t5.
DepDelay, t6.DepDelay_all,t7.ArrDelay,t8.ArrDelay_all,t9.Distance_origin,t10.
nDistance_origin,t11.Distance_dest,t12.nDistance_dest FROM airdata.t1,airdata.t2,
airdata.t3,airdata.t4,airdata.t5,airdata.t6, airdata.t7,airdata.t8,airdata.t9,
airdata.t10,airdata.t11,airdata.t12 where t1.origin=t2.dest and t1.origin=t3.origin
and t1.origin=t4.dest and t1.origin=t5.origin and t1.origin=t6.origin
and t1.origin=t7.dest and t1.origin=t8.dest and t1.origin=t9.origin and
t1.origin=t10.origin and t1.origin=t11.dest and t1.origin=t12.dest;

###退出hive环境，在linux下将hive中的表格下载到本地，存储为clusterdata.csv
exit();

hive -S -e"select * from airdata.clusterdata" >clusterdata.csv
```

4.3.2 用 Spark 建立分类模型

上述处理好的 2000—2008 年起飞延误的数据中共有 48 485 553 个观测，其中起飞延误

比例为 43.6%. 依然随机选取 60% 作为训练集, 40% 作为测试集, 调用 Spark 中的 MLlib 的随机森林程序分析数据, 测试集结果见表 4.14 和表 4.15. 数据分析程序 d3–1.py, d3–2.py 和 d3–3. py如下所示.

```python
#! /usr/bin/python
#-*- coding:utf-8 -*-
import time

OUTPUT_BASE_DIR = '/home/liuyang/projects/air'

t0 = time.time()

dict_month_to_season = {
    '1': '3', '2': '3', '3': '0', '4': '0', '5': '0',
    '6': '1', '7': '1', '8': '1', '9': '2', '10': '2',
    '11': '2', '12': '3'
    }
dict_dayofweek_to_weekday = {'1': '0', '2': '0', '3': '0', '4': '1', '5': '1', '6': '2', '7': '3'}

def split_time(s):
    t = int(s)
    if t >= 0 and t < 500:
        return '0'
    elif t >= 500 and t < 800:
        return '1'
    elif t >= 800 and t < 1100:
        return '2'
    elif t >= 1100 and t < 1700:
        return '3'
    elif t >= 1700 and t < 21000:
        return '4'
    elif t >= 2100 and t < 24000:
        return '5'
    else:
        return -1

f = open(f'OUTPUT_BASE_DIR/classdata0008.csv', 'r') #d1.txt中创建的文件
w = open(f'OUTPUT_BASE_DIR/classdata_cleaned0008.csv', 'w')

i = 0
for line in f:
    if 'NULL'in line:
```

```python
            continue
        tmp = line.strip().split(' ')
        if line[0] == '-'or line[0] == '0':
            try:
                l = ','.join(['0'] + [dict_month_to_season[tmp[1]]] +
                    [dict_dayofweek_to_weekday[tmp[2]]] + [split_time(tmp[3])] +
                    [split_time(tmp[4])] + tmp[5:]
                    )
            except:
                continue
        else:
            try:
                l = ','.join(['1'] + [dict_month_to_season[tmp[1]]] +
                    [dict_dayofweek_to_weekday[tmp[2]]] + [split_time(tmp[3])] +
                        [split_time(tmp[4])] + tmp[5:]
                    )
            except:
                continue
        w.write(l + '\n')

        #if i == 500:
        # break
        #i += 1
f.close()
w.close()
print(time.time() - t0)
```

```python
#! /usr/bin/python
#-*-coding:utf-8-*-
import random

OUTPUT_BASE_DIR = '/home/liuyang/projects/air'

f = open(f'OUTPUT_BASE_DIR/classdata_cleaned0008.csv', 'r') #d1.txt中创建的文件
w_train = open(f'OUTPUT_BASE_DIR/classdata_cleaned0008_train.csv', 'w') #训练集
w_test = open(f'OUTPUT_BASE_DIR/classdata_cleaned0008_test.csv', 'w') #测试集
#i = 0
for line in f:
    if random.random() < 0.6:
        w_train.write(line)
    else:
        w_test.write(line)
    #i += 1
```

```
    #if i == 100:
    #    break
f.close()
w_train.close()
w_test.close()
```

```
#! /usr/bin/env python
from pyspark import SparkContext, SparkConf
from pyspark.mllib.regression import LabeledPoint
from pyspark.mllib.classification import LogisticRegressionWithLBFGS
from pyspark.mllib.tree import RandomForest
from pyspark.sql import SQLContext

OUTPUT_BASE_DIR = '/home/liuyang/projects/air'

# Configuration if you use spark-submit
conf = SparkConf().setAppName("Test Application")
conf = conf.setMaster("local[10]")
sc = SparkContext(conf=conf)
sqlCtx = SQLContext(sc)

def create_label_point(line):
    line=line.strip().split(',')
    #line = [0 if x == ''or x == '-'or x == 'NULL'else x for x in line]
    #depdelay = 1.0
    #if float(line[0]) == 0:
    #    depdelay = 0.0
    return LabeledPoint(line[0], [float(x) for x in line[1:]])
    #return LabeledPoint(depdelay, [float(x) for x in line[1:]])

train = sc.textFile(f"file://OUTPUT_BASE_DIR/classdata_cleaned0008_train.csv").
map(create_label_point)
test = sc.textFile(f"file://OUTPUT_BASE_DIR/classdata_cleaned0008_test.csv").
map(create_label_point)

print("rf start")
# Train a RandomForest model.
model = RandomForest.trainClassifier(
    train, numClasses=2, categoricalFeaturesInfo=0: 4,1: 4,2: 7,3: 7,
    numTrees=20, featureSubsetStrategy="auto", impurity="gini", maxDepth=5, maxBins=64
    )

# Evaluate model on test instances and compute test error
```

```
predictions = model.predict(test.map(lambda x:  x.features))
labels_and_preds = test.map(lambda p:  p.label).zip(predictions)
#testErr = labels_and_preds.filter(lambda v:  v[0] != v[1]).count() / float(test_size)
#print(f"Testing Error = testErr")

# Confusion Matrix
testErr_11 = labels_and_preds.filter(lambda v:  v == (1, 1)).count()
testErr_10 = labels_and_preds.filter(lambda v:  v == (1, 0)).count()
testErr_01 = labels_and_preds.filter(lambda v:  v == (0, 1)).count()
testErr_00 = labels_and_preds.filter(lambda v:  v == (0, 0)).count()

print(testErr_11)
print(testErr_10)
print(testErr_01)
print(testErr_00)

accuracy = (float(testErr_11) + float(testErr_00)) / (float(testErr_11) +
float(testErr_10) + float(testErr_01) + float(testErr_00))
recall = float(testErr_11) / (float(testErr_11) + float(testErr_10))
precision = float(testErr_11) / (float(testErr_11) + float(testErr_01))
F1_measure = 2 * precision * recall / (precision + recall)

print(f'accuracy:  accuracy')
print(f'recall:  recall')
print(f'precision:  precision')
print(f'F1_measure:  F1_measure')
```

表 4.14　Spark 中随机森林预测结果 (1)

真实标签	预测标签	
	0	1
0	9231780	1703696
1	5955064	2505770

表 4.15　Spark 中随机森林预测结果 (2)

	Accuracy	Recall	Precision	F_measure
RF(tree=100)	0.6051	0.2967	0.5953	0.3955

可以看出整体预测准确率不高, 只有 60.51%. 究其原因, 不区分特定航线的起飞延误这个问题较难预测. 它的影响因素远远多于单机版中只选择一条航线情况下的影响因素. 这说明并不是数据量越大越容易得到好的结果, 数据量增大反而增加了问题的难度. 这里我们只是抛砖引玉, 给出一个示例. 希望读者进一步尝试其他方法. 数据分析是一个开放的问题, 没有最好, 只有更好.

第 5 章
公共自行车数据案例分析

5.1 数据简介

本案例的数据有两部分: 第一部分是纽约市公共自行车的交易流水表. 公共自行车与共享单车不同, 不能使用手机扫码在任意地点借还车, 而需要使用固定的自行车车桩借还车. 第二部分是纽约市天气数据.

5.1.1 交易流水表

交易流水表指的是用户借还车的记录, 数据集可以从 citibike 官方网址 https://www.citibikenyc.com/system-data 下载, 也可以从中国人民大学出版社网站下载. 本案例的数据集包含 2013 年 7 月 1 日至 2016 年 8 月 31 日共 38 个月 (1158 天) 的数据, 每个月一个文件, 数据大小约为 6GB, 其中, 2013 年 7 月到 2014 年 8 月的数据格式与之后各月的数据格式有所差别, 具体体现为变量 start time 和 stop time 的存储格式不同, 前者以 YYYY-m-d HH:MM:SS 形式存储, 后者以表 5.1 中所示的方式存储, 各变量说明如表 5.1 所示.

表 5.1 公共自行车数据字段简介表

变量编号	变量名	变量含义	变量取值及说明
1	trip duration	旅行时长	骑行时间, 数值型, 秒
2	start time	出发时间	借车时间, 字符串, m/d/YYYY HH:MM:SS
3	stop time	结束时间	还车时间, 字符串, m/d/YYYY HH:MM:SS
4	start station id	借车站点编号	定性变量, 站点唯一编号
5	start station name	借车站点名称	字符串
6	start station latitude	借车站点纬度	数值型
7	start station longitude	借车站点经度	数值型
8	end station id	还车站点编号	定性变量, 站点唯一编号
9	end station name	还车站点名称	字符串
10	end station latitude	还车站点纬度	数值型
11	end station longitude	还车站点经度	数值型

续表

变量编号	变量名	变量含义	变量取值及说明
12	bike id	自行车编号	定性变量,自行车唯一编号
13	user type	用户类型	Subscriber: 年度用户; Customer:24 小时或 7 天的临时用户
14	birth year	出生年份	仅此列存在缺失值
15	gender	性别	0: 未知; 1: 男性; 2: 女性

5.1.2 纽约市天气数据

因为天气对自行车的使用情况会产生较大的影响,因此作者编写爬虫程序从 https://www.wunderground.com/history/airport/KNYC 下载了相应日期的数据,并存储在 weather_data_NYC.csv 文件中. 该文件包含 2010 年 1 月 1 日至 2016 年 11 月 30 日的小时级别的天气数据,读者可以自行选出需要的时间段的天气数据,天气数据的字段含义如表 5.2 所示. 读者可以编写程序自行下载,也可以从中国人民大学出版社网站下载. 本案例的所有数据和程序均可从中国人民大学出版社网站下载.

表 5.2 天气数据字段简介表

变量编号	变量名	变量含义	变量取值及说明
1	date	日期	字符串, YYYY-m-d
2	time	时间	EDT(Eastern Daylight Timing) 指美国东部夏令时间
3	temperature	气温	单位: 摄氏度
4	dew_point	露点①	单位: 摄氏度
5	humidity	湿度	百分数
6	pressure	海平面气压	单位: 百帕
7	visibility	能见度	单位: 千米
8	wind_direction	风向	离散型, 类别包括 west, calm 等
9	wind_speed	风速	单位: 千米每小时
10	moment_wind_speed	瞬间风速	单位: 千米每小时
11	precipitation	降水量	单位: 毫米, 存在缺失值
12	activity	活动②	离散型, 类别包括 snow 等
13	conditions	状态③	离散型, 类别包括 overcast, light snow 等
14	WindDirDegrees	风向角	连续型, 取值为 0 ~ 359
15	DateUTC	格林尼治时间	YYYY/m/d HH:MM

注: ① 露点, 又称露点温度, 气象学中指在固定气压下, 空气中所含的气态水达到饱和而凝结成液态水所需降至的温度.
② 活动 (activity) 指一天的天气.
③ 状态 (conditions) 指一个小时内具体的天气情况, 比如说, 纽约一天是中雨, 但某个时间段是小雨, 则这个时间段的记录为: 活动: 中雨, 状态: 小雨.

5.2 单机实现

5.2.1 描述统计分析与可视化展现

1. 自行车使用总量与站点数量的时间序列分析

使用程序 e1.py 进行数据预处理，e2.py 画图，可以绘制出月级别的借车总量随着时间变化的时序图，如图 5.1 所示，图中虚线代表有借车交易的站点数量 (右坐标轴)，实线代表每个月份的自行车使用量 (左坐标轴)。

```python
"""
该py文件首先根据原始数据生成需要的参数文件,包括data.txt,data_out.txt,count.csv和
station_num.csv文件
以及weather_cleaned.csv文件
"""

import os
import pandas as pd
import numpy as np

OUTPUT_BASE_DIR = '/home/liuyang/projects/newyork_bike'
DATA_BASE_DIR = '/home/opendata/new_york_bike'

os.chdir(OUTPUT_BASE_DIR)

dates = (['2013-0'+ str(i) for i in np.arange(7, 10)] + \
         ['2013-'+ str(i) for i in np.arange(10, 13)] + \
         ['2014-0'+ str(i) for i in np.arange(1, 10)] + \
         ['2014-'+ str(i) for i in np.arange(10, 13)] + \
         ['2015-0'+ str(i) for i in np.arange(1, 10)] + \
         ['2015-'+ str(i) for i in np.arange(10, 13)] + \
         ['2016-0'+ str(i) for i in np.arange(1, 9)])
#生成data.txt文件
table_names = os.listdir(DATA_BASE_DIR)
table_names.remove('code')
table_names.remove('weather_data_NYC.csv')

with open('data.txt', 'w') as f:
    f.write('table_name\n')
```

```python
    for table in table_names:
        f.write(f'table\n')

#生成data_out.txt文件
with open('data_out.txt', 'w') as f:
    f.write('output_name\n')
    for date in dates:
        f.write(date.replace('-', '') + '\n')

#生成count.csv和station_num.csv文件
table_names = list(pd.read_table('data.txt')['table_name'])
count = []
stationnum = []
for table in table_names:
    data = pd.read_csv(f"DATA_BASE_DIR/table")
    count.append(len(data))
    stationnum.append(len(np.unique(data[['start station id','end station id']])))

pd.DataFrame(count, columns=['count']).to_csv('count.csv', index=False)
pd.DataFrame(stationnum, columns=['station number']).to_csv('station_num.csv', index=False)
names = [
    'date', 'time', 'temperature', 'dew_point', 'humidity', 'pressure', 'visibility',
    'wind_dir', 'wind_speed', 'wind_speed_in', 'precipitation', 'act', 'conditions',
    'WindDirDegrees', 'DateUTC'
    ]

weather_total = pd.read_csv(f'DATA_BASE_DIR/weather_data_NYC.csv')
weather_total.columns = names
weather_total['year'] = weather_total['date'].apply(lambda date: int(date.strip().split('-')[0]))
weather_total['month'] = weather_total['date'].apply(lambda date: int(date.strip().split('-')[1]))
weather_total['day'] = weather_total['date'].apply(lambda date: int(date.strip().split('-')[2]))
weather_total = weather_total[(weather_total['year'] >= 2013) & (weather_total['year'] <= 2016)]

names1 = ['date', 'time', 'temperature', 'dew_point', 'humidity', 'pressure', 'visibility',
        'wind_speed', 'conditions', 'year', 'month', 'day', 'hour']
names2 = ['date', 'time', 'temperature', 'dew_point', 'wet', 'pressure',
        'visibility', 'wind_speed', 'conditions', 'year', 'month', 'day', 'hour']
```

```python
weather_total['Date_time'] = weather_total['DateUTC'].apply(lambda time:
time.strip().split()[1])
weather_total['Date_hour'] = weather_total['Date_time'].apply(lambda time:
int(time.strip().split(':')[0]))
weather_total = weather_total[ weather_total[['year', 'month', 'day',
'Date_hour']].duplicated()]
weather_total['hour'] = (weather_total['Date_hour'] + 19)
weather_cleaned = weather_total[names1]
weather_cleaned.columns = names2
weather_cleaned.to_csv('weather_cleaned.csv', index=False)
```

```python
"""
该py文件画出图5.1
"""
import os
import pandas as pd
import matplotlib.pyplot as plt
import numpy as np

OUTPUT_BASE_DIR = '/home/liuyang/projects/newyork_bike'

os.chdir(OUTPUT_BASE_DIR)

#生成横坐标
dates = (['2013-0'+ str(i) for i in np.arange(7, 10)] + \
         ['2013-'+ str(i) for i in np.arange(10, 13)] + \
         ['2014-0'+ str(i) for i in np.arange(1, 10)] + \
         ['2014-'+ str(i) for i in np.arange(10, 13)] + \
         ['2015-0'+ str(i) for i in np.arange(1, 10)] + \
         ['2015-'+ str(i) for i in np.arange(10, 13)] + \
         ['2016-0'+ str(i) for i in np.arange(1, 9)])

#读取数据,注意数据要放到Python所在的工作路径,以后读取数据也要注意
count = pd.read_csv('count.csv')
station_num = pd.read_csv('station_num.csv')
#图5.1
fig = plt.figure()
plt.xticks(np.arange(len(dates))[1: : 2], dates[1: : 2], rotation='vertical')
ax1 = fig.add_subplot(1, 1, 1)
ax1.plot(count, 'k-', label='bike usage')
plt.legend(loc=2, bbox_to_anchor=(0, 0.85))
ax2 = ax1.twinx()
ax2.plot(station_num, 'k--', label='station num')
```

```
plt.legend(loc=2, bbox_to_anchor=(0, 0.75))
plt.title('Time Series of bike usage and station number')
plt.show()
fig.savefig('图5.1.png')
```

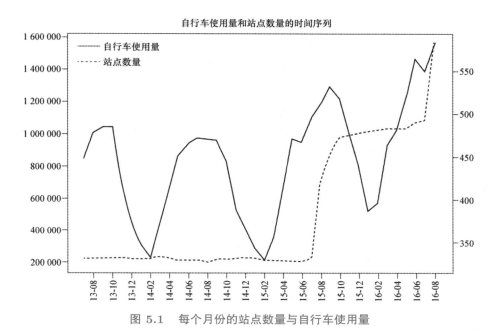

图 5.1　每个月份的站点数量与自行车使用量

从图 5.1 中可以看到在 2013 年 7 月至 2016 年 8 月这三年多的时间里, 自行车使用量有明显的季节性, 夏季借车总量明显高于冬季借车总量. 同时, 前两年的借车总量不包含趋势, 自行车站点的数量基本保持不变. 然而从 2015 年 8 月开始, 借车总量有了明显提升, 显著大于以前同期的借车总量, 这是因为纽约市的站点数有了明显上升, 具体数据是从 330 个站点增加到 419 个站点, 这一波站点的扩充持续到 2015 年 12 月, 最终维持在 472 个站点一直到 2016 年 7 月. 站点数从 2016 年 7 月开始又有一波扩充, 7 月和 8 月的站点数量分别为 483 个和 574 个.

2. 按节假日、周末和工作日划分

我们仅以 402 号站点为例展示节假日、周末和工作日的各个时间段内的平均借车次数, 感兴趣的读者可以尝试对不同站点进行分析. 对应程序为 e3.py 和 e4.py.

```
###请设置好工作目录，并将下面所需要的文件都放在工作目录下
setwd('/Users/hwh/Desktop/案例')
getwd()

library(colorspace)
```

```r
library(grid)
library(data.table)
library(VIM)
library(stringr)
library(dplyr)
library(ggplot2)

raw_rent_data<-read.csv('rent_data_402.csv',stringsAsFactors = F)
weather<-read.csv('weather_cleaned.csv',stringsAsFactors = F,sep = ',')

raw_rent_data<-merge(raw_rent_data,weather,all.x = T)
raw_rent_data<-raw_rent_data[order(raw_rent_data$year,raw_rent_data$month,raw_rent_data$day,raw_rent_data$hour),]
rownames(raw_rent_data)<-c(1:nrow(raw_rent_data))
raw_rent_data$date<-NULL
raw_rent_data$time<-NULL

data_v1<-raw_rent_data
data_v1$week<-factor(data_v1$week,levels = c('Monday','Tuesday','Wednesday',
'Thursday','Friday','Saturday','Sunday'))
mean(is.na(data_v1$conditions))
#缺失值比例约为0.43%，可以进行插补
set.seed(2032)
temp<-data_v1$temperature
data_v1$temperature[is.na(temp)]<-rnorm(sum(is.na(temp)),mean = mean(temp,na.rm = T),
                                       sd=sd(temp,na.rm = T))
temp<-data_v1$dew_point
data_v1$dew_point[is.na(temp)]<-rnorm(sum(is.na(temp)),mean = mean(temp,na.rm = T),
                                     sd=sd(temp,na.rm = T))
temp<-data_v1$humidity
data_v1$humidity[is.na(temp)]<-rnorm(sum(is.na(temp)),mean = mean(temp,na.rm = T),
                                    sd=sd(temp,na.rm = T))
temp<-data_v1$pressure
data_v1$pressure[is.na(temp)]<-rnorm(sum(is.na(temp)),mean = mean(temp,na.rm = T),
                                    sd=sd(temp,na.rm = T))
temp<-data_v1$visibility
data_v1$visibility[is.na(temp)]<-rnorm(sum(is.na(temp)),mean = mean(temp,na.rm = T),
                                      sd=sd(temp,na.rm = T))
data_v1$wind_speed[is.na(data_v1$wind_speed)]<-runif(sum(is.na(data_v1$wind_speed)),0,5)
data_v1$conditions[is.na(data_v1$conditions)]<-names(sort(table(data_v1$conditions),
decreasing = T)[1])
```

```r
aggr(data_v1)#无缺失值
#将必要的变量因子化:year,month,day,hour,week,conditions
for(variable in c('year','month','day','hour','week','conditions')){
  data_v1[,variable]<-as.factor(data_v1[,variable])
}

holidays<-read.csv('holidays.csv',header = F,sep = ',',stringsAsFactors = F)
colnames(holidays)<-c('date','holiday')
index<-holidays$holiday %in% c('Dr. Martin Luther King Jr.','Presidents Day',
                                'Veterans Day','Columbus Day')
holidays<-holidays[!index,]
holidays$year<-as.integer(str_sub(holidays$date,1,4))
holidays$month<-as.integer(str_sub(holidays$date,6,7))
holidays$day<-as.integer(str_sub(holidays$date,9,10))
holidays$date<-NULL

data_v2<-data_v1
data_v2<-merge(data_v2,holidays,all.x = T)
data_v2$holiday[!is.na(data_v2$holiday)]<-'Holiday'
index<-as.character(data_v2$week)%in%c('Saturday','Sunday')
data_v2$holiday[index]<-'Holiday'
data_v2$holiday[is.na(data_v2$holiday)]<-'No_Holiday'

data_v2$holiday<-as.factor(data_v2$holiday)
write.csv(data_v2,'data_v2.csv')
```

```python
# -*- coding: utf-8 -*-
"""
该py文件画出图5.2
"""
import pandas as pd
import matplotlib.pyplot as plt
import numpy as np

#读取R生成的数据
holidays = pd.read_csv('data_v2.csv')
classification = []
#进行分类,使得weekend用0表示,holiday用1表示,week用2表示
for week, holiday in zip(holidays['week'], holidays['holiday']):
    if week == 'Saturday'or week == 'Sunday':
        classification.append(0)
```

```
        elif holiday == 'Holiday':
            classification.append(1)
        else:
            classification.append(2)
holidays['classification'] = classification
holiday_group1 = holidays.groupby(['classification', 'hour'])
weekend_use, holiday_use, week_use = (holiday_group1['number'].mean().loc[0],
                                      holiday_group1['number'].mean().loc[1],
                                      holiday_group1['number'].mean().loc[2])
#图5.2
fig = plt.figure()
plt.plot(weekend_use, 'ok--', label='weekend usage')
plt.plot(holiday_use, '*k-', linewidth=2, label='holiday usage')
plt.plot(week_use, 'k-.', label='week usage')
plt.xticks(np.arange(0,24))
plt.xlim(0, 23)
plt.xlabel('time')
plt.ylabel('average usage')
plt.title('comparsion of weekend, holiday and week day')
plt.legend(loc=2)
plt.show()
fig.savefig('图5.2.png')
```

此处, 我们以美国法定节假日中的 New Years Day(1 月 1 日)、Memorial Day(5 月 27 日)、July 4th(7 月 4 日)、Labor Day(9 月 2 日)、Thanksgiving Day(11 月 28 日)、Christmas(12 月 25 日) 作为节假日, 周六日作为周末, 其余时间作为工作日进行划分, 得到一天中各个时间段 402 号站点的平均借车次数, 如图 5.2 所示.

由图 5.2 可以清楚地看出, 节假日各个时间段的平均借车次数几乎一致小于等于工作日的平均借车次数, 周末平均借车次数则介于二者之间, 并且周末平均借车次数呈单峰, 与其他时间段有明显不同.

3. 按天气划分

仍以 402 号站点为例分析雨天、雪天对自行车平均借出量的影响. 天气类型有多种分类, 如 "Clear" "Heavy Rain" "Light Rain" 等, 部分天气类型记录 (字段名:conditions) 存在缺失. 缺失比例大约为 0.43%, 这个比例很小, 为简单起见, 我们选取天气类型的众数 "Clear" 进行填补. 之后我们将类别中含有 "Rain" 或 "Snow" 字符串的时间段归为一类, 以此表示雨天、雪天, 其余天气作为另一类. 结果如图 5.3 所示 (程序见 e5.py).

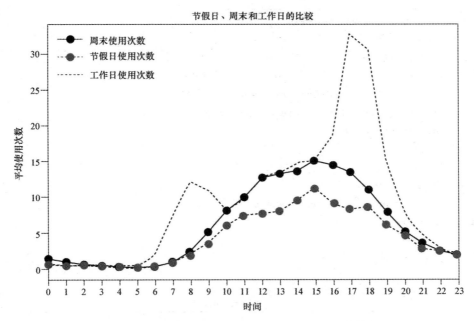

图 5.2　节假日、周末和工作日平均借车次数比较图

```
# -*- coding: utf-8 -*-
"""
该py文件用来作图5.3
"""
import pandas as pd
import matplotlib.pyplot as plt
import numpy as np

holidays = pd.read_csv('data_v2.csv')
#获取condition中带有Rain和Snow的
rain_or_snow = [condition for condition in np.unique(holidays.conditions)
                if 'Rain'in condition or 'Snow'in condition]

#对每个样本进行分类,对于雨天或雪天分类为1,否则为0
temp = []
for condition in holidays['conditions']:
    if condition in rain_or_snow:
        temp.append(1)
    else:
        temp.append(0)

holidays['rain or snow'] = temp
data_group2 = holidays.groupby(['rain or snow', 'hour'])
```

```
usage = data_group2['number'].mean()
other = usage[0]
rain_snow = usage[1]
fig = plt.figure()
plt.plot(other, 'k-.', linewidth=2, label='other weather')
plt.plot(rain_snow, 'ko-', label='rain or snow day')
plt.xticks(np.arange(0, 24))
plt.xlim(0, 23)
plt.xlabel('time')
plt.ylabel('average usage')
plt.title('comparsion of different weather')
plt.legend(loc=2)
plt.show()
fig.savefig('图5.3.png')
```

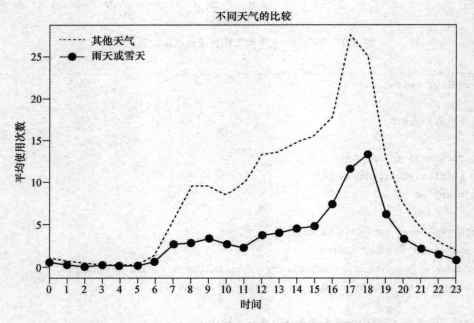

图 5.3　雨天或雪天与其他天气自行车平均使用量比较图

从图 5.3 中可以看出, 两条曲线有着相似的走势, 差别在于雨天或雪天的自行车平均使用量一直小于其他天气.

4. 网络可视化

(1) 动态气泡图. 利用 Python 绘制了本数据研究时期内每个月份的各站点自行车使用量气泡动画, 在此仅展示动画部分截图, 具体过程请参看源代码. 动态截图如图 5.4 所示. 对应程序是 e6.py,e7.py,e8.py.

```python
# -*- coding: utf-8 -*-
"""
该py文件用来生成站点经纬度坐标文件stationloc.csv和站点编号文件stationid.csv
以及每个站点借车量文件,命名格式为yyyymmstationcount.csv
"""
import pandas as pd

minputs = pd.read_table('data.txt')
moutputs = pd.read_table('data_out.txt')
stationloc = dict()
for minput, moutput in zip(minputs['table_name'], moutputs['output_name']):
    temp = pd.read_csv(minput)
    stationcount = temp.groupby('start station id')['tripduration'].count()
    #生成每个站点借车量文件
    stationcount.to_csv(str(moutput) + 'stationcount.csv')
    for ID, Lat, Lon in zip(
            temp['start station id'], temp['start station latitude'], temp['start station longitude']
            ):
        if ID not in stationloc.keys():
            stationloc[ID] = (Lon, Lat)

#生成站点经纬度坐标
stationloc = pd.DataFrame(
    list(stationloc.values()), index=stationloc.keys(), columns=['longitude', 'latitude']
    )
stationloc.to_csv('stationloc.csv')

stationloc = pd.read_csv('stationloc.csv', header=0, names=['id', 'longitude', 'latitude'])
stationloc['id'].to_csv('stationid.csv', index=False)
```

```python
# -*- coding: utf-8 -*-
"""
该py文件为画出动态气泡图做准备,将每一帧图像保存在一张图片内为将来的动态气泡图做准备
"""
import pandas as pd
import matplotlib.pyplot as plt
import numpy as np
import os
```

```python
stationloc = pd.read_csv('stationloc.csv', header=0, names=['id', 'longitude',
'latitude'])

#station字典里存储每个站点的经纬度
station = dict()
for loc, x, y in np.array(stationloc):
    if int(loc) not in station.keys():
        station[int(loc)] = (x, y)

#读取当前路径下的站点借车量文件
filelist = os.listdir(os.getcwd())
check_files = []
for f in filelist:
    if 'stationcount.csv'in f:
        check_files.append(f)
check_files.sort()

#作出动态图5.4所需图像
N = 1
fig = plt.figure(figsize=(4, 19 * 2))
for minput in check_files:
    fig.add_subplot(19, 2, N)
    stationcount = pd.read_csv(minput, header=None)
    stationcount = np.array(stationcount)
    station_x = []
    station_y = []
    for i, j in stationcount:
        if i in station.keys():
            station_x.append(station[i][0])
            station_y.append(station[i][1])
    plt.ylim(40.66, 40.82)
    plt.xlim(-74.06, -73.92)
    plt.scatter(
        x=station_x,
        y=station_y,
        edgecolors='k',
        color='k',
        s=150 * stationcount[:, 1] / float(max(stationcount[:, 1]))
        )
    plt.xticks([])
    plt.yticks([])
    plt.box(False)
    plt.subplots_adjust(wspace=0, left=0, bottom=0, right=1, top=1, hspace=0)
```

```
        N += 1
fig.savefig('station.png')
"""
该py代码是本案例中唯一使用Python3编写的代码,该动态图也可以利用其他模块编写,请读者自行
试验(gif图)
"""

import itertools, sys, time, random, math, pygame
from pygame.locals import *

class MySprite(pygame.sprite.Sprite):
    def __init__(self,target):
        pygame.sprite.Sprite.__init__(self)
        self.master_image = None
        self.frame = 0
        self.old_frame = -1
        self.frame_width = 1
        self.frame_height = 1
        self.first_frame = 0
        self.last_frame = 0
        self.columns = 1
        self.last_time = 0

    def _getx(self):
        return self.rect.x

    def _setx(self,value):
        self.rect.x = value

    X = property(_getx, _setx)

    def _gety(self):
        return self.rect.y
    def _sety(self,value):
        self.rect.y = value

    Y = property(_gety, _sety)

    def _getpos(self):
        return self.rect.topleft

    def _setpos(self,pos):
        self.rect.topleft = pos
```

```
        position = property(_getpos, _setpos)

    def load(self, filename, width, height, columns):
        self.master_image = pygame.image.load(filename).convert_alpha()
        self.frame_width = width
        self.frame_height = height
        self.rect = Rect(0, 0, width, height)
        self.columns = columns
        rect = self.master_image.get_rect()
        self.last_frame = (rect.width // width) * (rect.height // height) - 1

    def update(self, current_time, rate=30):
        if current_time > self.last_time + rate:
            self.frame += 1
            if self.frame > self.last_frame:
                self.frame = self.first_frame
            self.last_time = current_time
        if self.frame != self.old_frame:
            frame_x = (self.frame % self.columns) * self.frame_width
            frame_y = (self.frame // self.columns) * self.frame_height
            rect = Rect(frame_x, frame_y, self.frame_width, self.frame_height)
            self.image = self.master_image.subsurface(rect)
            self.old_frame = self.frame

    def __str__(self):
        return (
            str(self.frame) + ','+ str(self.first_frame) + ','+ str(self.last_frame) +
            ','+ str(self.frame_width) + ','+ str(self.frame_height) + ','+ str(self.columns) +
            ','+ str(self.rect)
            )

def print_text(font, x, y, text, color=(255, 255, 255)):
    imgText = font.render(text, True, color)
    screen.blit(imgText, (x, y))

station13=['2013-0'+ str(i) for i in range(7, 10)] + ['2013-'+ str(i) for i in range(10, 13)]
station14=['2014-0'+ str(i) for i in range(1, 10)] + ['2014-'+ str(i) for i in range(10, 13)]
station15=['2015-0'+ str(i) for i in range(1, 10)] + ['2015-'+ str(i) for i in range(10, 13)]
```

```python
station16=['2016-0'+ str(i) for i in range(1, 10)]
station = station13 + station14 + station15 + station16
pygame.init()
screen = pygame.display.set_mode((144, 144), 0, 32)
pygame.display.set_caption('station animation')
font = pygame.font.Font(None, 18)
Font = pygame.font.SysFont('simhei', 18)
framerate = pygame.time.Clock()
animation = MySprite(screen)
animation.load('station.png', 144, 144, 2)
group = pygame.sprite.Group()
group.add(animation)
running = True
i = 0
while running:
    framerate.tick(1)
    ticks = pygame.time.get_ticks()
    for event in pygame.event.get():
        if event.type == pygame.QUIT:
            running = False
    keys = pygame.key.get_pressed()
    if keys[pygame.K_ESCAPE]:
        running = False
        screen.fill((0, 0, 100))
        group.update(ticks)
        group.draw(screen)
        i += 1
        if i < len(station):
            print_text(font, 10, 10, station[i], (0, 0, 100))
        else:
            i = 0
            print_text(font, 10, 10, station[i], (0, 0, 100))
    pygame.display.update()
pygame.quit()
```

图 5.4(a) 和 (b) 分别展示了 2013 年 7 月和 2016 年 8 月各个站点借车总量的气泡图, 图中气泡越大, 表示该站点自行车借出量越大. 从动态图中可以明显地看出不同时间站点借车量的变化以及站点数量的变化.

(2) 网络图分析. 以 2016 年 8 月 3 日为例进行网络图分析, 如图 5.5(a) 所示, 感兴趣的读者可以对其他时间的网络进行分析. 我们绘制了各个站点在这一天的自行车借还情况, 该网络图是有向图, 箭头从借车站点指向还车站点 (很多站点之间同时有借还记录, 所以大部分站点两两之间是双向连接), 图中点越大, 颜色越亮, 表明该站点的借还车总量越大. 可以看出,

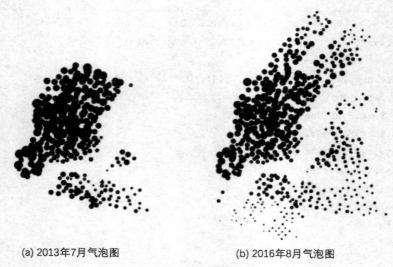

(a) 2013年7月气泡图　　　　(b) 2016年8月气泡图

图 5.4　动态气泡图截图

在一天的时间内发生了多次借车交易, 庞大的点数和边数使我们无法看清细节, 因此选取图中黑色方框圈住的部分进行描述分析 (黑色方框圈住的部分表示经度位于 $40.695° \sim 40.72°$, 纬度位于 $-74.023° \sim -73.973°$ 之间的区域), 如图 5.5(b) 所示. 对应程序为 e9.py.

```
# -*- coding: utf-8 -*-
"""
该py文件对2016年8月的自行车使用情况进行网络图分析,画出图5.5
"""
import pandas as pd
import matplotlib.pyplot as plt
import numpy as np
import networkx as nx
import matplotlib.patches as patches

stationloc = pd.read_csv('stationloc.csv', header=0, names=['id', 'longitude',
'latitude'])

station = dict()
for loc, x, y in np.array(stationloc):
    if int(loc) not in station.keys():
        station[int(loc)] = (x, y)

data1608 = pd.read_csv('201608-citibike-tripdata.csv')
starttime = data1608['starttime']
starttime.to_csv('starttime.csv')
#选取2016年8月1日进行网络图分析#
```

```python
startday = []
startindex = []
starthour = []
with open('starttime.csv') as f:
    while True:
        line = f.readline()
        if line:
            line = line.strip().split(',')
            day = line[1].split('')[0].split('/')[1]
            hour = line[1].split('')[1].split(':')[0]
            startindex.append(float(line[0]))
            startday.append(day)
            if day == '3':
                starthour.append(hour)
        else:
            break
startday = np.array(startday)
startindex = np.array(startindex, dtype=np.long)
starthour = np.array(starthour)
unihour = np.unique(starthour)
day3 = startindex[np.where(startday == '3')[0]]#取出8月3日的行标#
G3 = nx.DiGraph()
record = np.array(data1608[['start station id', 'end station id']])
for start,end in record[day3]:
    G3.add_edge(start, end)
node_size = [float(G3.degree(v)) for v in G3]
fig = plt.figure(figsize=(6, 10))
nx.draw(G3, pos=station, node_size=node_size, node_color=node_size, edgecolor='k')
#截取一部分点进行分析#

fig1 = plt.figure(figsize=(6, 10))
ax1 = fig1.add_subplot(1, 1, 1, aspect='equal')
plt.ylim(40.60, 40.85)
plt.xlim(-74.04, -73.90)
ax1.add_patch(patches.Rectangle((-74.023, 40.695), 0.05, 0.025, fill=False,
edgecolor='k', linewidth=3))

nx.draw(G3, pos=station, ax=ax1, node_size=node_size, node_color=node_size,
edgecolor='k')
plt.show()

station_condition = dict()
```

```python
for loc, x, y in np.array(stationloc):
    if x > -74.023 and x < -73.973 and y > 40.695 and y < 40.72:
        if int(loc) not in station_condition.keys():
            station_condition[int(loc)] = (x, y)
station_not_include = list(station.keys())
for loc in station_condition.keys():
    station_not_include.remove(loc)
G3_remove = nx.DiGraph(G3)
G3_remove.remove_nodes_from(station_not_include)
fig2 = plt.figure(figsize=(6, 10))
ax2 = fig2.add_subplot(1, 1, 1, aspect='equal')
node_size_remove = [float(G3_remove.degree(v)) for v in G3_remove]
nx.draw(G3_remove, pos=station, node_size=node_size_remove, node_color=node_size_remove)
#使用算法使网络更美观#

#网络分析
path = nx.all_pairs_shortest_path(G3_remove)
nx.average_shortest_path_length(G3_remove) # 网络平均最短距离
degree = nx.degree(G3_remove)
closenesss = nx.closeness_centrality(G3_remove)
betweenness = nx.betweenness_centrality(G3_remove)
G3_remove.number_of_nodes() # 节点数量
G3_remove.number_of_edges() # 连接数量
nx.density(G3_remove) # 网络密度
nx.transitivity(G3_remove) ##传递性
nx.diameter(G3_remove)
```

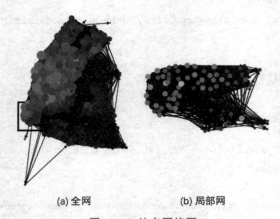

(a) 全网　　　　　　(b) 局部网

图 5.5　站点网络图

对图 5.5(b) 进行网络基本特征分析, 结果展示在表 5.3 中.

表 5.3　局部网描述统计分析

节点数	边数	网络密度	平均最短路径长度	网络直径
80	2066	0.327	1.738	4

表 5.3 描述了选定区域的网络的基本特征,该网络共有 80 个节点,2066 条边,网络密度 0.327 表示边的个数占所有可能的连接数的比例,如果比例较低,则说明点之间的连接并不完全,即大部分的点没有直接相连. 在此分析的网络图是连通图,即网络内的节点都是可达的. 平均最短路径长度表示两点之间平均最少需要经过接近两条边才能相连,网络直径是 4 表示顶点之间最多需要经过 4 个顶点就可以相互连接. 由上述统计量可以看出,这幅图中顶点之间的联系是比较密切的.

(3) 选取特定两个站点进行分析. 图 5.6 是从 2006 号站点 (Central Park S & 6 Ave) 到 3143 号站点 (5 Ave & E 78 St) 的自行车线路. 这是一条热门线路,A,B,C (左、中、右) 代表的是谷歌地图返回的三条可选骑行路线,图像及数据的获取由 R 语言包 ggplot2 中的相关函数完成. 感兴趣的读者可以参看源代码进行练习. 对应程序为 e10.R.

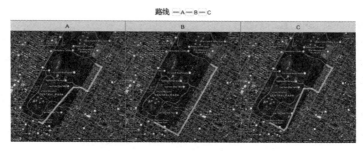

图 5.6　三条可选骑行路线图

```
work_dir<-getwd()
work_dir='/Users/hwh/Desktop/案例'
setwd(work_dir)
getwd()

###为了防止下面地图的绘制出现错误,请安装ggmap的dev版本
library(devtools)
devtools::install_github("dkahle/ggmap")

library(dplyr)
library(VIM)
library(ggplot2)
library(ggmap)
library(stringr)
library(lubridate)
```

```r
library(reshape2)

###201607-citibike-tripdata.csv文件已经存储在all_tables文件夹中
nyc<-read.csv('201607-citibike-tripdata.csv',stringsAsFactors = F)
colnames(nyc)
colnames(nyc)<-c('tripduration','start_time','end_time','start_id',
                 'start_name','start_lat','start_lon','end_id',
                 'end_name','end_lat','end_lon','bike_id','usertype',
                 'birth_year','gender')

stations<-nyc[!duplicated(select(nyc,end_id)),
              c('end_id','end_lon','end_lat','end_name')]
colnames(stations)<-c('id','lon','lat','name')
ggplot(stations,mapping = aes(x=lon,y=lat))+geom_point()
#异常站点,经纬度错误
filter(stations,lon>-60)
stations<-filter(stations,lon<(-60))
stations<-stations[order(stations$id),]
ggplot(stations,mapping = aes(x=lon,y=lat))+geom_point()
#进一步去除偏远站点
stations1<-filter(stations,lon>(-74.02),lat>40.675)
stations1<-stations1[order(stations1$id),]
ggplot(stations1,mapping = aes(x=lon,y=lat))+geom_point()

#获取曼哈顿地区的地图
(max(stations1$lon)+min(stations1$lon))/2;(max(stations1$lat)+min(stations1$lat))/2
mh_map<-get_map(location = c(lon=-73.97282,lat=40.73987),zoom = 12)
ggmap(mh_map)
ggmap(mh_map)+geom_point(data=stations1[c('lon','lat')],color='blue')
ggmap(mh_map)+geom_point(data=stations[c('lon','lat')],color='blue')

#将站点的名称从原始数据集删除
nyc<-select(nyc,-start_name,-end_name)
#去除异常站点所在记录
nyc<-filter(nyc,start_lon<(-60),end_lon<(-60))
#对时间进行转换
nyc['start_time']<-apply(nyc['start_time'],MARGIN = 2 ,
                         FUN = strptime,format='%m/%d/%Y %H:%M:%S')
nyc['end_time']<-apply(nyc['end_time'],MARGIN = 2 ,
                       FUN = strptime,format='%m/%d/%Y %H:%M:%S')
nyc$start_time<-as.POSIXct(nyc$start_time)
nyc$end_time<-as.POSIXct(nyc$end_time)
```

```
nyc$start_year<-year(nyc$start_time)
nyc$start_month<-month(nyc$start_time)
nyc$start_day<-day(nyc$start_time)
nyc$start_hour<-hour(nyc$start_time)
nyc$start_min<-minute(nyc$start_time)
nyc$start_sec<-second(nyc$start_time)
nyc$start_week<-weekdays(nyc$start_time,abbreviate = F)

nyc$end_year<-year(nyc$end_time)
nyc$end_month<-month(nyc$end_time)
nyc$end_day<-day(nyc$end_time)
nyc$end_hour<-hour(nyc$end_time)
nyc$end_min<-minute(nyc$end_time)
nyc$end_sec<-second(nyc$end_time)
nyc$end_week<-weekdays(nyc$end_time,abbreviate = F)

#借车热度
grouped1<-group_by(nyc,start_id)
heat_of_borrow<-summarise(grouped1,n=n())
heat_of_borrow[,c('start_lon','start_lat')]<-
filter(stations,id%in%(heat_of_borrow$start_id))[,c('lon','lat')]
heat_of_borrow$n<-heat_of_borrow$n/31
MHD_Map<-ggmap(mh_map,extent = 'device',legend = 'topleft')
MHD_Map+geom_point(aes(x=start_lon,y=start_lat,size=n/max(n),
                    color='purple',alpha=n/max(n)),
                data=heat_of_borrow)

#还车热度
grouped2<-group_by(nyc,end_id)
heat_of_return<-summarise(grouped2,n=n())
heat_of_return[,c('end_lon','end_lat')]<-
filter(stations,id%in%(heat_of_return$end_id))[,c('lon','lat')]
heat_of_return$n<-heat_of_return$n/31
MHD_Map+geom_point(aes(x=end_lon,y=end_lat,size=n/max(n),
                    color='purple',alpha=n/max(n)),
                data=heat_of_return)

#热门借还路线
grouped3<-group_by(nyc,start_id,end_id)
diff_route<-summarise(grouped3,n=n())
diff_route<-diff_route[order(diff_route$n,decreasing = T),]
head(diff_route)
#选取2006,3143站点进行研究
```

```
filter(stations,id%in%c(2006,3143))

#不知为何，route函数不能正确获取5 Ave & E 78 St, New York的位置
#将5 Ave & E 78 St, New York换成其附近的一个地点Ukrainian Institute of America
my_legs1<-route('Central Park S & 6 Ave, New York',
                'Ukrainian Institute of America, New York',
                mode='bicycling',alternatives = T)
my_legs2<-route('Central Park S & 6 Ave, New York',
                'Ukrainian Institute of America, New York',
                mode='bicycling',alternatives = F)
#热门线路附近的底图
my_basemap<-qmap('Rumsey Playfield, East 71st Street, New York, NY, United States',
                 zoom = 15, maptype = 'hybrid',
                 base_layer = ggplot(aes(x = startLon, y = startLat),
                                     data = my_legs2))
#绘制三条线路
my_basemap+
  geom_leg(aes(
    x=startLon,y=startLat,
    xend=endLon,yend=endLat,color=route),
    alpha=3/4,size=2,data=my_legs1
  )+labs(x='Lon',y='Lat',color='Route')+facet_wrap( route,ncol=3)+theme(legend.position =
'top')
```

5.2.2 自行车角度的分析

1. 自行车使用情况描述统计分析

经过统计分析，对应程序为 e11.py．

```
# -*- coding: utf-8 -*-
"""
该部分从自行车角度分析
该py文件生成数据文件,包括bikecount.csv,biketime.csv,bikestation.csv
文件bikecount.csv每行代表一个自行车ID,每列表示从2013年7月1日起每天自行车的使用次数
文件biketime.csv每行代表一个自行车ID,每列表示从2013年7月1日起每天自行车的使用时长(单位
为s)
文件bikestation.csv每行代表一个自行车ID,每列代表一个站点,矩阵中的元素a_ij表示第i辆自行
车出现在第j个站点的次数
"""
import pandas as pd
```

```python
def bike_info(file_names):
    biketime = dict()
    bikecount = dict()
    bikestation = dict()
    for file_name in file_names:
        with open(file_name, 'r') as f:
            f.readline()
            while True:
                line = f.readline()
                if line:
                    line = line.strip().split(',')
                    day = line[1].split('')[0].replace('"', '')
                    ID = line[11].replace('"', '')
                    time = float(line[0].replace('"', ''))
                    start = line[3].replace('"', '')
                    end = line[7].replace('"', '')
                    if ID in biketime.keys():
                        if day in biketime[ID].keys():
                            biketime[ID][day] += time
                            bikecount[ID][day] += 1
                        else:
                            biketime[ID][day] = time
                            bikecount[ID][day] = 1
                        if start in bikestation[ID].keys():
                            bikestation[ID][start] += 1
                        else:
                            bikestation[ID][start] = 1
                        if end in bikestation[ID].keys():
                            bikestation[ID][end] += 1
                        else:
                            bikestation[ID][end] = 1
                    else:
                        biketime[ID] = dict()
                        biketime[ID][day] = time
                        bikecount[ID] = dict()
                        bikecount[ID][day] = 1
                        bikestation[ID] = dict()
                        bikestation[ID][start] = 1
                        if start == end:
                            bikestation[ID][start] += 1
                        else:
                            bikestation[ID][end] = 1
                else:
```

```python
                        break
    return bikecount, biketime, bikestation

def save_dict(result, name):
    keys = result.keys()
    g = open(f'name.csv', 'w')
    for key in keys:
        g.write(str(key) + ','+ ','.join([str(i) for i in result[key]]) + '\n')
    g.close()


minputs = pd.read_table('data.txt')
inputs = list(minputs['table_name'])
days = []
for f in inputs:
    tmp = pd.read_csv(f)
    for time in list(tmp['starttime']):
        strings = time.strip().split('')[0]
        if strings not in days:
            days.append(strings)
pd.DataFrame(days, columns=['days']).to_csv('days.csv')
#记录每辆自行车每天的骑行次数以及骑行时间
days = list(pd.read_csv('days.csv', index_col=0)['days'])
stationid = list(pd.read_csv('stationid.csv', header=None, names=['stationid'])
['stationid'])
bikecount, biketime, bikestation = bike_info(inputs)
res_count = dict()
res_time = dict()
res_station = dict()
for ID in bikecount.keys():
    res_count[ID] = []
    res_time[ID] = []
    res_station[ID] = []
    for day in days:
        if day in bikecount[ID].keys():
            res_count[ID].append(bikecount[ID][day])
            res_time[ID].append(biketime[ID][day])
            else:
            res_count[ID].append(0)
            res_time[ID].append(0)
        for Id in stationid:
            Id = str(Id)
            if Id in bikestation[ID].keys():
                res_station[ID].append(bikestation[ID][Id])
```

```
            else:
                res_station[ID].append(0)
save_dict(res_count, 'bikecount')
save_dict(res_time, 'biketime')
save_dict(res_station, 'bikestation')
```

在本研究时间段共 1158 天内共 11487 辆自行车有使用记录, 3878 辆自行车在数据收集第一天 (2013 年 7 月 1 日) 开始使用, 7295 辆自行车在数据收集结束时 (2016 年 8 月 31 日) 仍在使用. 自行车使用情况的描述统计量如表 5.4 所示. 其中, 总天数表示每辆自行车在研究期间使用的天数, 从表中可以看出, 有些自行车在研究期间只在某一天使用过, 有的自行车的使用天数为 1154 天, 接近研究长度 1158 天; 总次数表示在研究期间每辆自行车使用的次数; 总时长表示在研究期间每辆自行车使用的总时间, 单位为小时; 平均每天骑行时长用骑行总时长除以实际使用天数得到; 平均每天骑行次数用骑行总次数除以实际使用天数得到. 从表中的平均每天骑行时长的分布情况可以看出, 有很少一部分自行车的平均每天骑行时长十分异常, 回归到原始数据, 我们发现这些异常车辆的骑行次数非常少, 有的仅 1 次, 但是骑行时间却非常长, 有一辆自行车的骑行时间甚至超过了纽约规定的自行车最长使用时间 (24 小时). 在后续分析中, 将删除这一部分自行车.

表 5.4　自行车使用情况的描述分析

名称	统计量	最小值	25 分位点	中位数	平均数	75 分位点	最大值
Day	总天数 (天)	1	252	883	685.45	1153	1154
Count	总次数 (次)	1	1290.5	3070	2757.04	4209	5450
Time	总时长 (小时)	0.028	318.51	769.84	714.57	1083.04	3574.37
Average time	平均每天骑行时长 (小时)	0.0063	0.9253	1.0443	1.391	1.684	106.345
Average count	平均每天骑行次数 (次)	0.0234	3.644	4.034	5.364	6.935	26

注: 总天数用自行车末次使用日期减去初次使用日期 +1 来表示.

程序 e12.py 对上述结果进行了画图 (见图 5.7).

```
# -*- coding: utf-8 -*-
"""
该py文件用来画图5.7
"""
import pandas as pd
import matplotlib.pyplot as plt
import numpy as np

bikecount = pd.read_csv('bikecount.csv', header=None, index_col=0)
biketime = pd.read_csv('biketime.csv', header=None , index_col=0)
bikeid = np.array(bikecount.index)
start = []
end = []
```

```python
bikecount = np.array(bikecount)
N, p = bikecount.shape
for i in np.arange(N):
    bikeappear = np.where(bikecount[i] != 0)[0]
    start.append(bikeappear[0])
    end.append(bikeappear[-1])

start = np.array(start)
end = np.array(end)
Index = np.argsort(start)[: : -1]
id_sort = bikeid[Index]
start_sort = np.array(start)[Index]
end_sort = np.array(end)[Index]
start_uniques = np.unique(start_sort)[: : -1]

#对相同起始日期的自行车再按照终止日期进行排序
sort_index = []
for start_unique in start_uniques:
    tmp_index = np.arange(N)[np.array(start) == start_unique]
    tmp_end = end[tmp_index]
    tmp = np.argsort(tmp_end)[: : -1]
    sort_index += list(tmp_index[tmp])

sort_index = sort_index[: : -1]

fig = plt.figure()
plt.plot(np.arange(N), start[sort_index], linewidth=5, color='k')
plt.plot(np.arange(N), end[sort_index], 'k')
plt.xticks(np.arange(0, N, 1185), id_sort[sort_index][: : 1185], rotation=45)
plt.ylabel('days')
plt.xlabel('bike id')
plt.title('start day and end day')
fig.savefig('图5.7(a).png')

fig = plt.figure()
plt.plot(np.arange(5000), start[sort_index][: 5000], linewidth=5, color='k')
plt.plot(np.arange(5000), end[sort_index][: 5000], 'k')
plt.xticks(np.arange(0, 5000, 1185)[: 5], id_sort[sort_index][: 5000: 1185], rotation=45)
plt.ylabel('days')
plt.xlabel('bike id')
plt.title('start day and end day')
fig.savefig('图5.7(b).png')
```

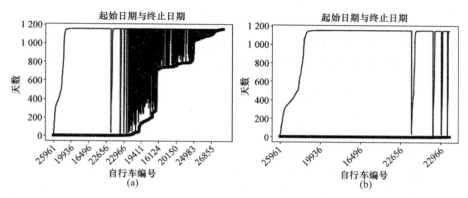

图 5.7 自行车起始终止日期

图 5.7(a) 展示了每辆自行车起始终止日期轮廓图, 每个横坐标对应一辆自行车, 纵坐标表示时间, 0 代表 2013 年 7 月 1 日, 1157 代表 2016 年 8 月 31 日. 对于每辆自行车, 画出其对应的纵坐标的两个点, 分别是起始日期点和终止日期点, 首先按照起始日期升序排序, 然后对同样起始日期自行车的终止日期进行升序排序, 最后将自行车的起始日期和终止日期分别连线, 就得到了自行车起始终止日期轮廓图. 为了更清晰地看出自行车使用情况, 我们将最早使用的 5000 辆自行车取出进行展示, 如图 5.7(b) 所示.

2. K 均值聚类

在此我们对自行车进行 K 均值聚类分析, 程序为 e13.py.

```
# -*- coding: utf-8 -*-
"""
该py代码对经过处理的数据进行K均值聚类分析
"""
import pandas as pd
import matplotlib.pyplot as plt
import numpy as np
from sklearn.preprocessing import StandardScaler
from sklearn.cluster import KMeans


bikecount = pd.read_csv('bikecount.csv', header=None, index_col=0)
biketime = pd.read_csv('biketime.csv', header=None , index_col=0)
bikeid = np.array(bikecount.index)
total_days = bikecount.apply(
    lambda record:  np.where(record != 0)[0][-1] - np.where(record != 0)[0][0] + 1,
axis=1
    )
total_counts = bikecount.sum(axis=1)
total_times = biketime.sum(axis=1) / 3600
```

```
#构造变量
avg_count = np.array(total_counts / total_days)
avg_time = np.array(total_times / total_days)
#去除异常值
avg_time_clean = avg_time[avg_time <= 8]
avg_count_clean = avg_count[avg_time <= 8]

#将数据进行标准化
X = np.hstack((avg_count_clean.reshape(-1, 1), avg_time_clean.reshape(-1, 1)))
x = StandardScaler().fit_transform(X)

kmeans_model = KMeans(n_clusters=3).fit(x)
labels = kmeans_model.labels_
label0 = (labels == 0)
label1 = (labels == 1)
label2 = (labels == 2)
fig = plt.figure()
ax = fig.add_subplot(1, 1, 1)
plt.plot(x[:, 0][label0], x[:, 1][label0], 'k*', label='label0')
plt.plot(x[:, 0][label1], x[:, 1][label1], 'k.', label='label1')
plt.plot(x[:, 0][label2], x[:, 1][label2], 'ko', label='label2')
plt.xlabel('average count')
plt.ylabel('average time')
plt.legend()
fig.savefig('图5.8.png')

labelcenter = kmeans_model.cluster_centers_
```

选用表 5.4 中列出的平均每天骑行时长和平均每天骑行次数作为变量进行聚类分析. 通过对变量的分析发现, 数据中存在一些异常值, 这些异常值对于 K 均值聚类造成不良影响, 因此我们决定去除这些异常点, 以后再进行聚类分析. 把平均每天骑行时长超过 8 小时 (占比少于 0.1%) 的样本点作为异常点, 共去掉 12 辆异常自行车, 利用标准化以后的变量对剩余样本进行聚类. 将自行车聚为 3 类, 聚类结果如图 5.8 所示, 三类自行车的聚类中心和数量如表 5.5 所示.

从表 5.5 中可以看出, 第三类自行车属于使用频繁、使用时长较长的自行车; 第一类自行车与第三类自行车正好相反, 属于使用不够频繁、使用时长较短的自行车. 这两类自行车可以通过适当调度、合理使用, 减少维修成本, 延长使用寿命. 有 2047 辆自行车属于第二类自行车, 这些自行车的使用时长以及使用次数都接近整体平均水平.

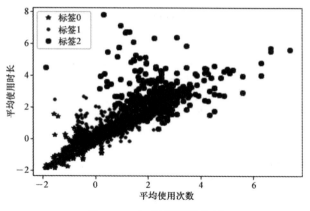

图 5.8 K 均值聚类结果

表 5.5 K 均值聚类中心

	第一类 (星)	第二类 (小圆点)	第三类 (大圆点)
平均使用次数	−0.578	0.728	2.035
平均使用时长	−0.556	0.624	2.062
类内样本数	7910	2047	1518

3. 站点交叉 (双向聚类)

我们按自行车在站点的出现次数将所有自行车分为经常出现和不经常出现两种. 如果自行车在该站点的出现次数大于全部自行车在该站点出现次数的中位数, 则认为此自行车经常出现在该站点, 取值为 1, 否则取值为 0. 在此我们选取首次使用日期距离数据收集起始日期 700 ∼ 800 天、末次使用日期距离数据收集起始日期 1000 天以上的 1797 辆自行车, 去除这些自行车从未使用过的部分站点, 用 1797 辆自行车对剩余的 616 个站点进行双向聚类. 图 5.9(a) 展示了原始数据情况, 图中颜色深的小方格 (数字是 1) 表示该辆自行车在该站点经常出现. 从图中可以看出, 图中不同颜色的小方块分布比较散乱, 我们尝试使用双向聚类, 将取值为 1 的小方块放在一起. 在此展示 CC 算法聚类的结果 (见图 5.9(b)). 对于 CC 算法聚类, 根据分类效果展示图, 将数据聚成 16 类, 使得每一行或者每一列只属于其中一类.

图 5.9(b) 展示了 CC 算法双向聚类的结果, 程序为 e14.py.

```
# -*- coding: utf-8 -*-
"""
该py文件用来进行双向聚类,作图5.9(b)
"""
import pandas as pd
import matplotlib.pyplot as plt
import numpy as np
from sklearn.cluster import SpectralCoclustering
```

```python
bikecount = pd.read_csv('bikecount.csv', header=None, index_col=0)
bikeid = np.array(bikecount.index)
start = []
end = []
bikecount = np.array(bikecount)
N, p = bikecount.shape
for i in np.arange(N):
    bikeappear = np.where(bikecount[i] != 0)[0]
    start.append(bikeappear[0])
    end.append(bikeappear[-1])

start = np.array(start)
end = np.array(end)
bikestation = pd.read_csv('bikestation.csv', header=None, index_col=0)
condition = (start > 700) & (start < 800) & (end > 1000)
id_conditions = bikeid[condition]
bike_station = np.array(bikestation)[condition]
bike_station = pd.DataFrame(bike_station)
bike_station_freq = pd.DataFrame()
for column in list(bike_station.columns):
    bike_station_freq[column] = np.array(bike_station[column] >
bike_station[column].median(), dtype=int)

INDEX0 = np.ones(bike_station_freq.shape[0], dtype=bool)
INDEX0[[np.where(bike_station_freq.sum(axis=1) == 0)[0]]] = False
INDEX1 = np.ones(bike_station_freq.shape[1], dtype=bool)
INDEX1[[np.where(bike_station_freq.sum(axis=0) == 0)[0]]] = False
bike_station_freq_clean = np.array(bike_station_freq)[INDEX0, :][:, INDEX1]

#画出原始数据矩阵图
fig = plt.figure(figsize=(5, 10))
ax = fig.add_subplot(1, 1, 1)
ax.matshow(bike_station_freq_clean, cmap=plt.cm.Blues)
#画出CC算法得到的结果图
for i in range(2, 21):
    model = SpectralCoclustering(n_clusters=i, random_state=0)
    model.fit(bike_station_freq_clean)
    fit_data = np.array(bike_station_freq_clean)[np.argsort(model.row_labels_)]
    fit_data = fit_data[:, np.argsort(model.column_labels_)]
    fig = plt.figure(figsize=(5, 10))
    ax = fig.add_subplot(1, 1, 1)
    ax.matshow(fit_data, cmap=plt.cm.Blues)
```

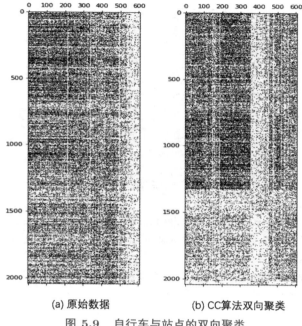

图 5.9 自行车与站点的双向聚类

我们展示的是将自行车和站点分类结果进行适当顺序调整后的结果. 可以看出, 颜色相近的小方块相比原始数据来说更多聚集在一起, 大体将自行车和站点进行了分类. CC 算法将自行车和站点分成 16 类, 但在图中较为明显的是左上角和右下角两类, 其余类的样本量都比较小. 在这两类样本块中, 左上角的样本块中包含 1 068 辆自行车以及 320 个站点; 右下角的样本块中包含 468 辆自行车以及 201 个站点, 这些自行车编号和站点编号不再给出, 可以在 Python 中查看. 图中其余的样本块由于自行车和站点数量较少而不太明显.

读者也可以对该双向聚类进行扩展分析, 可以结合前一部分的 K 均值聚类来分析站点与自行车之间的联系, 比如, 考察高频使用站点是否与高频使用自行车有更密切的关系; 也可以利用双向聚类的结果再对自行车进行 K 均值聚类分析, 更深层次的探索请读者自行尝试.

5.2.3 单个站点借车量预测分析

下面对单个站点公共自行车的借车量进行预测. 我们要研究的问题是, 通过站点过去一段时间的借车量和天气数据来预测未来的单日借车量.

以 402 号站点为例. 统计得到 402 号站点每天的自行车借用量时间序列数据, 将 2016 年 8 月 18 日 (不包括 18 日) 前的数据作为训练集, 8 月 18—31 日的数据作为测试集. 程序见 e15.py.

```
"""该py文件实现三阶指数平滑模型"""

import matplotlib.pyplot as plt
```

```python
import pandas as pd
import numpy as np
import math
import itertools
from sklearn import linear_model

class HoltWinters(object):

    """
    基于网格搜寻最优值的Holt-Winters算法,借鉴github作者etlundquist编写的HoltWinters算
    法,加上自己的改进编写而成。
        网址:https://github.com/etlundquist/holtwint
    该类以数据挖掘思想作为背景,将时间序列划分为训练集和测试集,目的是为了寻找到合适的参
    数alpha, beta和gamma。
        算法以MAPE作为评价标准,使用者也可以自定义评价指标,对应修改类方法Compute_Mape即可,
    方法GridSearch也需要进行修改。
        @params:
            - ts:   时间序列(序列时间由远及近)
            - p[int]:   时间序列的周期
            - test_num[int]:   测试集长度
            - sp[int]:   计算初始化参事所需要的周期数(周期数必须大于1)
            - ahead[int]:   需要预测的滞后数
            - mtype[string]:   时间序列方法:累加法或累乘法['additive'/'multiplicative']
    """

    ............

if __name__ == '__main__':
    data = pd.read_csv('rent_data_402.csv')
    number = data['number']
    days = data[['year', 'month', 'day']]\
                .apply(
                    lambda dte: '{}/{}/{}'.format(dte[0], dte[1], dte[2]),
                    axis=1
                    )
    number = data.groupby(['year', 'month', 'day'])['number'].sum()
    number.index = days.unique()
    number = number.astype('float')

    HWmodel = HoltWinters(number, 365, 14)
```

```
    HWmodel.GridSearch()

    plt.plot(np.array(HWmodel.ts_test), 'ko-', label='True Values')
    plt.plot(HWmodel.best_pred, 'k.-', label='fitted values')
    plt.legend()
    plt.xticks(np.arange(len(HWmodel.ts_test))[: : 2], HWmodel.ts_test.index[: :
2], rotation=315)
    plt.show()
    plt.savefig('图5.11.png')
```

1. 时间序列预测

首先, 我们尝试对日级别的自行车借用量进行时间序列预测分析, 画出时间序列数据的时序图, 如图 5.10 所示.

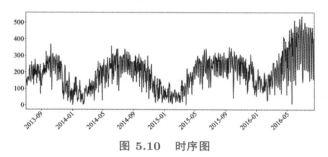

图 5.10　时序图

从图 5.10 中可以看出, 自行车借用量数据存在明显的周期性和趋势性, 由于三阶指数平滑可以较好地拟合具有周期性和趋势性的时间序列, 因此选择三阶指数平滑模型对数据进行建模. 利用 $\text{MAPE} = \frac{1}{n}\sum_{i=1}^{n}\frac{|y_i - \hat{y}_i|}{y_i} \cdot 100\%$ 评价指标来选取最优参数 alpha, beta 和 gamma. 利用 8 月 18 日前的数据建立模型, 采取网格搜寻策略寻找预测效果最优的模型, 以预测该站点 8 月 18 日以后的自行车借用量, 将预测值和真实值展示在图 5.11 中.

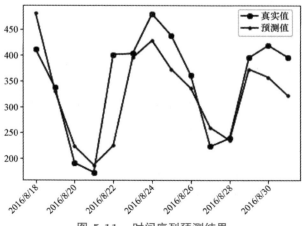

图 5.11　时间序列预测结果

从图 5.11 中可以看出, 三阶指数平滑模型的预测效果一般, 计算得到 MAPE 为 12.89%, 因此, 将增加额外信息 (自变量) 建立随机森林模型来对借车量进行更准确的预测.

2. 随机森林

时间序列预测中整理好的 402 号站点的日级别的自行车借用量, 加上天气数据就构成了原始数据集, 数据集中共包含 11 个变量, 其中变量 number 是因变量, 自变量包括 year, month, day, week, temperature, dew_point, humidity, pressure, visibility, wind_speed 共 10 个. 我们将 2016 年 8 月 18 日 (不包括 18 日) 前的数据作为训练集, 8 月 18—31 日的数据作为测试集.

现在, 尝试建立随机森林模型来预测公共自行车的借用量. 利用 Python 程序 e16.py 构建拥有 2~50 棵树的随机森林模型, 利用训练集 (2016 年 8 月 4 日前的数据) 训练模型, 利用选择集 (2016 年 8 月 4—17 日的数据) 选择模型, 选出最优模型以后, 利用训练集和选择集中全部样本训练模型, 最后利用测试集 (2016 年 8 月 18—31 日的数据) 评价模型效果, 利用 MAPE 作为评价指标. 其中, 每个随机森林模型重复进行 5 次试验, 最后将 5 次试验结果的平均值作为这个随机森林模型最终的 MAPE. 平均 MAPE 结果如图 5.12 所示.

```python
# -*- coding: utf-8 -*-
"""
该py文件建立随机森林模型对借车量进行预测
"""
import pandas as pd
import matplotlib.pyplot as plt
import numpy as np
from sklearn.ensemble import RandomForestRegressor

data1 = pd.read_csv('rent_data_402.csv')
#生成param.csv文件
param = data1[['year', 'month', 'day', 'week', 'hour']].drop_duplicates()
param.to_csv('param.csv', index=False)

#随机森林预测
number = data1['number']

days = data1[['year', 'month', 'day']] \
        .apply(
            lambda dte: '{}/{}/{}'.format(dte[0], dte[1], dte[2]),
            axis=1
        )
```

```python
number = data1.groupby(['year', 'month', 'day'])['number'].sum()
number.index = days.unique()
number = number.astype('float')
xticks = number[-14:].index

data2 = pd.read_csv('weather_cleaned.csv')
data = pd.merge(data1, data2, on=['year', 'month', 'day', 'hour'])
data_clean = data[['year', 'month', 'day', 'week']].drop_duplicates()
data_clean.index = np.arange(len(data_clean))
group_data = data.groupby(['year', 'month', 'day']).mean()
group_data = group_data.drop('hour', 1)
group_data.index = np.arange(len(group_data))
group_data = group_data.drop('number', 1)
#对每天的数据进行预测分析
dis_var = ['year', 'month', 'day', 'week']
xnames = list(data_clean.columns)
dis_x = pd.DataFrame(index=np.arange(len(data_clean)))
for var in dis_var:
    xnames.remove(var)
    dis_x = dis_x.join(pd.get_dummies(data_clean[var], prefix=var))
y = number
y.index = np.arange(len(number))

x = group_data.join(dis_x)
conditions = (data_clean['year'] == 2016) & (data_clean['month'] == 8) & (data_clean['day'] >= 18)

train_x = x[: -conditions.sum()]
test_x = x[-conditions.sum():]
train_y = y[: -conditions.sum()]
test_y = y[-conditions.sum():]
pd.DataFrame(train_y).join(train_x).to_csv('pre_train.csv', index=False, header=None)
pd.DataFrame(test_y).join(test_x).to_csv('pre_test.csv', index=False, header=None)

train = pd.read_csv('pre_train.csv', header=None)
test = pd.read_csv('pre_test.csv', header=None)

x_train, y_train = train.drop(0, axis=1), train[0]
x_test, y_test = test.drop(0, axis=1), test[0]

def Mape(y_true, y_pred):
    y_true = np.array(y_true, dtype=float)
    y_pred = np.array(y_pred)
```

```python
        return(np.mean(np.abs(y_true - y_pred) / y_true) * 100)

MAPE = []
for n_estimator in np.arange(2, 51):
    mape = 0
    for _ in np.arange(5):
        rf = RandomForestRegressor(n_estimators=n_estimator, random_state=0)
        rf.fit(x_train[: -14], y_train[: -14])
        y_predict = rf.predict(x_train[-14:])
        mape += Mape(y_train[-14:], y_predict)
    MAPE.append(mape / 5.)

fig = plt.figure()
plt.plot(np.arange(2, 51), MAPE, 'k')
plt.xticks(np.arange(2, 51, 2))
plt.xlabel('ntrees')
plt.ylabel('average MAPE')
plt.title('Different RandomForest Models')
plt.show()
fig.savefig('图5.12.png')
rf = RandomForestRegressor(n_estimators=np.arange(2, 51)[np.argmin(MAPE)])
rf.fit(x_train, y_train)
y_predict = rf.predict(x_test)
fig = plt.figure(figsize=(10, 5))
ax = fig.add_subplot(111)
plt.plot(np.array(y_test), 'ko-', label='true values')
plt.plot(y_predict, 'k.-', label='fitted values')
plt.legend()
plt.xticks(np.arange(len(xticks))[: : 2], xticks[: : 2], rotation=315)
plt.show()
fig.savefig('图5.13.png')
```

由图 5.12 可以看出,随着回归树数量的增加,MAPE 呈现下降趋势,当树的数量达到 7 时, MAPE 开始出现反弹. 因此建立 7 棵树的随机森林模型对借车量进行预测,画出预测值和真实值的折线图,如图 5.13 所示,预测的 MAPE 为 8.22%.

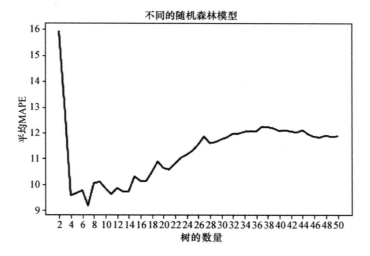

图 5.12　随机森林的平均 MAPE

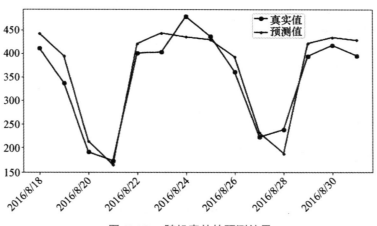

图 5.13　随机森林的预测结果

5.3　分布式实现

5.3.1　数据预处理与描述统计

我们利用 Hive 进行数据预处理. 仅以实现图 5.1 的数据预处理为例, 给出利用 Hive 对数据进行预处理的步骤以及具体语句, 程序为 f1.hql.

```
自行车案例图5.1数据准备:
1.首先,输入hive进入到hive中;
```

2.通过命令SHOW DATABASES;可以查看hive中现存的数据库(注意不要丢掉分号,必须每个命令都带有分号);

3.通过命令CREATE DATABASE IF NOT EXISTS bikedata;可以创建bikedata数据库;

4.通过命令USE bikedata;可以进入刚创建好的数据库;

5.通过命令SHOW TABLES;可以查看数据库中现存的表格,需要注意的是hive数据库每次只能存入一张表格,如果需要存入30张表格就必须运行30次命令;

6.通过下列命令建表

CREATE TABLE IF NOT EXISTS 201307bikedata(
tripduration string,
starttime string,
stoptime string,
start_station_id string,
start_station_name string,
start_station_latitude string,
start_station_longitude string,
end_station_id string,
end_station_name string,
end_station_latitude string,
end_station_longitude string,
bikeid string,
usertype string,
birth_year string,
gender string)
ROW FORMAT DELIMITED FIELDS TERMINATED BY ','
STORED AS TEXTFILE;

7.将数据导入对应的表格中

LOAD DATA LOCAL INPATH '/home/opendata/new_york_bike/2013-07 - Citi Bike trip data.csv' OVERWRITE INTO TABLE 201307bikedata;

8.若要创建结构完全相同的表格,可以通过命令

CREATE TABLE IF NOT EXISTS 201308bikedata like 201307bikedata;

来创建表格,然后再将数据导入新创建的表格中,由于过程都一致,所以我们仅以一个表格作为例子,去寻找图5.1中位于2013年7月的自行车使用量以及站点数;

9.获得2013年7月的自行车使用量:

由于第一行导入的是表的列名,并不算做一次记录,因此要用总行数减1;由此,可以获得该时间段内的自行车使用量

SELECT count(*)-1 FROM 201307bikedata;

对于其他时间段自行车的使用量也可以类似得到,将每个月使用量汇总即可用于绘制图5.1中的实线;

10.通过命令

SELECT start_station_id FROM 201307bikedata UNION DISTINCT SELECT end_station_id FROM 201307bikedata;

可以获得两列的并集,之后通过计数就可以求得两列共有多少元素,即该时间段内共出现多少不重复的自行车站点;

```
通过命令
SELECT count(*)-2 FROM
    (SELECT start_station_id FROM 201307bikedata
        UNION DISTINCT
     SELECT end_station_id FROM 201307bikedata)
    station_num;
```
这里的减2是减去了两列的列名

至此,我们获得了2013年7月这一段时间内共出现的不重复自行车站点数量,运行相同的命令在不同的数据集中即可获得图5.1中虚线所需要的数据,在此不再赘述

5.3.2 分布式预测模型

我们利用 Spark 对所有有效站点的借车总量进行预测, 仍然以 2016 年 8 月 18 日 (不包含 18 日) 前的数据作为训练集训练模型,2016 年 8 月 18 日以后的数据作为测试集进行测试. 程序为 f2.py 和 f3.py.

```python
# -*- coding: utf-8 -*-
"""
该py文件用来生成每个站点在整个数据收集期间的自行车借用数据,用来在Spark上对全部站点进行
随机森林预测
"""
import os
import pandas as pd

DATA_BASE_DIR = '/home/opendata/new_york_bike'

#生成rent_data_站点id.csv
file_paths = list(pd.read_table('data.txt')['table_name'])

dataset = dict()
for file_path in file_paths:
    with open(os.path.join(DATA_BASE_DIR, file_path), 'r') as f:
        f.readline()
        while True:
            line = f.readline().replace('"', '').strip().split(',')
            if line:
                date_time = line[1]
                date, time = date_time.strip().split(' ')
                if '-' in date:
                    year, month, day = date.strip().split('-')
                else:
                    month, day, year = date.strip().split('/')
```

```
                hour = time.strip().split(':')[0]
                station_id = int(line[3])
                record = pd.DataFrame(index=[0])
                record['year'], record['month'], record['day'] = int(year), int(month),
int(day)
                record['hour'], record['number'] = int(hour), 1
                if station_id in dataset.keys():
                    dataset[station_id] = dataset[station_id].append(record)
                else:
                    dataset[station_id] = record
            else:
                break
param = pd.read_csv('param.csv')
for station_id in dataset.keys():
    tmp_data = dataset[station_id]
    data = param.merge(tmp_data, on=['year', 'month', 'day', 'hour'], how='left')
    data.fillna(0, inplace=True)
    data.groupby(['year', 'month', 'day', 'week', 'hour']).sum().to_csv(f'rent_data_{station_id}.csv')
"""
该py文件将来会运行在pyspark中,用来对所有满足要求的站点借车量进行随机森林预测
"""
from pyspark import SparkContext, SparkConf
from pyspark.mllib.regression import LabeledPoint
from pyspark.mllib.tree import RandomForest
import pandas as pd
import os
import numpy as np

# Configuration if you use spark-submit
conf = SparkConf().setAppName("randomforest").setMaster("local")
sc = SparkContext(conf=conf)
weather = pd.read_csv('weather_cleaned.csv')

file_lists = list(pd.read_table('rent_data.txt', sep=',', header=None)[0])
dataset = []
pred_0 = []
no_pred = []
pred_id = []

for file_name in file_lists:
    tmp_data = pd.read_csv(os.path.join(os.getcwd(), file_name))
```

```python
        data = pd.merge(tmp_data, weather, on=['year', 'month', 'day', 'hour'])
        data_clean = data[['year', 'month', 'day', 'week']].drop_duplicates()
        data_clean.index = np.arange(len(data_clean))
        group_data = data.groupby(['year', 'month', 'day']).mean()
        group_data = group_data.drop('hour', 1)
        group_data.index = np.arange(len(group_data))
        group_data = group_data.drop('number', 1)
        data_clean = data_clean.join(group_data)
        data_clean['number'] = np.array(data.groupby(['year', 'month',
'day']).sum()['number'])
        try:
            train_index = np.where(data_clean['number'][: -14] != 0)[0]
            first_day, last_day = train_index[0], train_index[-1]
            train_total_day = last_day - first_day + 1
            if last_day < len(data_clean) - 14 * 2:
                pred_0.append(file_name)
            elif train_total_day >= 7:
                pred_id.append(file_name)
                dataset.append(data_clean[first_day:])
            else:
                no_pred.append(file_name)
        except:
            pass

#按照要求将站点名分配到不同文件,将来只对pred_id.txt文件中的文件进行预测
with open('pred_0.txt', 'w') as f:
    for i in pred_0:
        f.write(f'{i}\n')

with open('no_pred.txt', 'w') as f:
    for i in no_pred:
        f.write(f'{i}\n')

with open('pred_id.txt', 'w') as f:
    for i in pred_id:
        f.write(f'{i}\n')

def run_rf_spark(data):
    try:
        dis_var = ['year', 'month', 'day', 'week']
        xnames = list(data.columns)
        xnames.remove('number')
```

```python
        def replace(dataframe, columns):
            for column in columns:
                trep = dict()
                for key, value in zip(dataframe[column].unique(),
np.arange(len(dataframe[column].unique()))):
                    trep[key] = value
                dataframe.loc[:, column] = dataframe.loc[:, column].replace(trep)
            return dataframe

    x = data[xnames]
    y = data['number']
    x = replace(x, dis_var)
    param_dict = {}
    dis_loc = [0, 1, 2, 3]
    for loc, column in zip(dis_loc, dis_var):
        param_dict[loc] = len(x[column].unique())
    train_x = x[:  -14]
    test_x = x[-14:]
    train_y = y[:  -14]
    test_y = y[-14:]
    test_var = np.mean((test_y - test_y.mean()) ** 2)

def get_train(line):
    return LabeledPoint(line[0], line[1:])

def get_test(line):
    return line[1:]

def get_y(line):
    return line[0]

train = sc.parallelize(np.array(pd.DataFrame(train_y).join(train_x)))
test = sc.parallelize(np.array(pd.DataFrame(test_y).join(test_x)))
train_spark = train.map(get_train)
test_spark_x = test.map(get_test)
test_spark_y = test.map(get_y).collect()
if test_var > 0.00001:
    nt = list(range(1, 10)) + list(range(10, 50, 5))
    best_MSE = np.inf
    for t in nt:
        for _ in range(10):
            model_rf = RandomForest.trainRegressor(train_spark, param_dict, t)
            predictions_rf = model_rf.predict(test_spark_x).collect()
```

```
                MSE = np.mean((np.array(predictions_rf) - np.array(test_spark_y)) **
2)
                if MSE < best_MSE:
                    best_predict = predictions_rf
                    best_MSE = MSE
        else:
            best_MSE = -1
            best_predict = []
        return [best_MSE] + list(best_predict)
    except:
        return [-1]
#利用循环进行预测,由于服务器限制无法对全部站点同时进行预测,如果同时进行预测会产生数据
量过大,导致spark无法运行
for file_name, data in zip(pred_id, dataset):
    result_mllib = pd.DataFrame(run_rf_spark(data))
    result_mllib.to_csv(f'result_{file_name}', index=False)
```

在进行站点借车量预测时需要注意的是, 在这 613 个站点中, 有些站点是数据收集期间增加的, 有些站点则在数据收集期间关闭. 由于随机森林模型需要一定数量的数据对模型进行训练才能获得较好的结果, 因此要求训练集的样本量尽量大. 训练集的有效天数小于 7 天的站点共有 64 个, 将这些站点视为无效站点, 不对其进行预测. 有 40 个站点在 8 月 4—17 日内的借车量为 0, 因此认为该部分站点被关闭, 且关闭状态一直延续到未来.

对剩余的 509 个站点的借车量进行预测, 在此采用与单机随机森林相似的策略选择最优模型, 选取决策树数量为 $1,2,\cdots,10,15,20,\cdots,45$ 的随机森林模型, 并且对每个随机森林模型进行 10 次训练, 这里采用均方误差 MSE 作为评价指标, 选出平均均方误差最小的模型在测试集上进行预测. 我们随机选取了 2 个站点以及 402 号站点的预测结果进行展示, 如图 5.14 所示.

在图 5.14 中, 小圆点线表示真实值, 大圆点线表示预测值. 从图中可以看出, 483 号站点的预测效果较差, 预测曲线偏离真实曲线较远, 可能是因为测试集中存在异常点, 从 483 号站点在测试集中的借车量趋势可以看出, 该站点第 11 天 (计数从 0 开始) 的自行车借车量突然变得很少, 实际数字为 5, 很可能是某些异常情况所致, 随机森林模型并不能从数据中捕获到这种异常情况, 因此, 导致均方误差和 MAPE 较大. 其他两个站点的随机森林预测效果比较令人满意, 三个站点的均方误差分别是 61.02, 1235.39 和 2742.88, MAPE 分别是 35.94%, 35.65%, 252.06%. 感兴趣的读者可以参看源代码对所有站点的借车量进行预测.

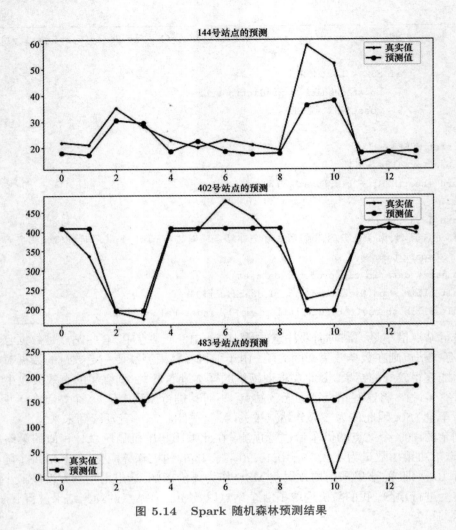

图 5.14 Spark 随机森林预测结果

第 6 章
机器翻译实例

6.1 数据简介与数据预处理

本章数据来自联合国官方平行语料（中英文文档）(http://opus.nlpl.eu/)，原始训练集数据量约 1 589 万条，测试集有 4 000 条，均为中英文对应语料. 在本案例中，由于 GPU 计算资源的限制，选取训练集前 30 万条数据进行建模.

6.1.1 删除异常值

首先对总体数据进行去重处理，即除去中英文均相同的数据，剩下 275 955 条. 对于中文语料文本，除去非中文行，如 "A"，或者只有编号、符号的，如图 6.1 所示. 非中文行共有 2 251 行，对应的英文语料也去除相应的行.

```
----不包含中文  21   A
----不包含中文  36   B
----不包含中文  47   C
----不包含中文  355  注
----不包含中文  358  ORIGINAL: ENGLISH
----不包含中文  2674 ORIGINAL: ENGLISH/SPANISH
----不包含中文  4283 男
----不包含中文  4298 女
----不包含中文  4358 n. d.
----不包含中文  4553 INE。
----不包含中文  4684 IEP。
----不包含中文  5294 ORIGINAL: SPANISH
----不包含中文  7093 CERD/C/287
----不包含中文  7736 A/AC. 109/1194
----不包含中文  7739 A/AC. 109/1198
----不包含中文  7740 A/AC. 109/1195
----不包含中文  7743 A/AC. 109/1197
----不包含中文  7749 (CERD/C/263/Add. 4)
----不包含中文  7751 (CERD/C/247/Add. 1)
----不包含中文  7754 (CERD/C/295/Add. 1)
```

图 6.1 删去中文语料中的非中文行

6.1.2 修改异常值及数据筛选

除去文本本来的空格,如 "目 录" 修改为 "目录", 主要为了防止后续分词出现异常. 此外, 对中文语句进行简繁体识别, 结果训练集中的中文语句均为简体.

考虑到长 (难) 句翻译在日常生活中的需求, 因此第二步是筛选出训练集中长度大于 50 的中文句子及其对应的英文句子. 经过筛选, 训练集保留 90 572 条句子, 验证集 3 000 条句子. 对于测试集采用同样的标准进行选取, 测试集数据量为 1 000 条. 使用本书 2.2.1 节介绍的 Seq2Seq 及 Transform 模型对上述数据建模实现英译中. 有兴趣的读者可以自行尝试中译英. 以上数据预处理程序见 preprocess.py.

```
#CLEAN::数据清理
import pandas as pd
import os
print(os.getcwd())
N_test = 5000

###############训练集###############
## 初步清理
# 取出长句子
file_zh = open("/rym/en-zh/train.zh")
file_en = open("/rym/en-zh/train.en")
zhs = []
ens = []
count = 0
N = 10000
while count < N:
    line_cn = file_zh.readline()
    line_en = file_en.readline()
    if len(line_cn) > 50:
        zhs.append(line_cn)
        ens.append(line_en)
        count += 1

# 验证对齐
n = 10000-1
print(ens[n])
print(zhs[n])

df = pd.DataFrame('en':ens, 'zh':zhs)
```

```python
# 去除句末换行符
for i in range(10000):
    df.zh[i] = df.zh[i].replace('\n', '')

df['en'] = df.apply(lambda x:  x['en'].replace('\n', ''), axis=1)
df['zh'] = df.apply(lambda x:  x['zh'].replace('\n', ''), axis=1)
df_copy = df.copy()

# 去重
print(len(set(df['en'])), len(set(df['zh'])))
df = df.drop_duplicates()
df.to_csv("/rym/en-zh/seq_tmp1.csv", index = False)

# 去除异常值
from collections import Counter

# 判断字符串是否包含中文
def is_contains_chinese(strs):
    for _char in strs:
        if '\u4e00'<= _char <= '\u9fa5':
            return True
    return False


def isvalid_zh(idx, string):
    if not isinstance(string, str):
        print("----不是字符串: ", idx, string)
        return False
    if len(string) < 2 or not is_contains_chinese(string):
        print("----不包含中文", idx, string)
        return False
#     if re.search(r"\W", string):
#         print("----有特殊字符", idx, string)
    return True

zhstrange_idx = [isvalid_zh(i, s) for i,s in enumerate(df['zh'])]
# 查看有多少条中文结果异常
Counter(zhstrange_idx)

def isvalid_en(idx, string):
    if not isinstance(string, str):
        print("----不是字符串: ", idx, string)
```

```
            return False
#     if is_contains_chinese(string):
#         print("----包含中文：", idx, string)
#         return False
    return True

enstrange_idx = [isvalid_en(i, s) for i,s in enumerate(df['en'])]
# 查看有多少条英文结果异常
Counter(enstrange_idx)
# 只要中文或英文异常即剔除
print(df.shape)
valid_index = [enstrange_idx[i] and zhstrange_idx[i] for i in range(df.shape[0])]
df = df[valid_index]
print(df.shape)
df.to_csv("/rym/en-zh/seq_clean1.csv", index = False)


## 深度清理
ens = list(df['en'])
zhs = list(df['zh'])
del df

# 英文部分
import nltk
nltk.download('punkt')
import re
replacement_patterns = [
    (r'\'', '&apos;'),
    (r'"', '&quot;')
]
class RegexpReplacer(object):
    def __init__(self, patterns=replacement_patterns):
        self.patterns = [(re.compile(regex), repl) for (regex, repl) in patterns]
    def replace(self, text):
        s = text
        for (pattern, repl) in self.patterns:
            s = re.sub(pattern, repl, s)
        return s

replacer = RegexpReplacer()
def en_process(ens):
    ens_new = []
```

```python
        count = 0
        for en in ens:
            count += 1
            if(count%(10**5) == 0):
                print(count)
            en_tokens = nltk.word_tokenize(en)
            en_space = ''.join(en_tokens)
            en = replacer.replace(en_space)
            ens_new.append(en)
        return(ens_new)

ens = en_process(ens)
N_test = 5000
ens_train = ens[:-N_test]
ens_test = ens[-N_test:]
print(len(ens), len(ens_train), len(ens_test))

f = open("/rym/en-zh/test_en.en","w")
for l in ens_test:
        f.write(l)
        f.write('\n')
f.close()
f = open("/rym/en-zh/train_en.en","w")
for l in ens_train:
        f.write(l)
        f.write('\n')
f.close()

# 中文部分
import jieba
def zh_process(zhs):
    zhs_new = []
    count = 0
    for zh in zhs:
        count += 1
        if(count%(10**5) == 0):
            print(count)
        zh = zh.replace(" ", "")
        zh_tokens = list(jieba.cut(zh))
        zhs_new.append(''.join(zh_tokens))
    return(zhs_new)
```

```python
zhs = zh_process(zhs)

N_test = 5000
zhs_train = zhs[:-N_test]
zhs_test = zhs[-N_test:]
print(len(zhs), len(zhs_train), len(zhs_test))

f = open("/rym/en-zh/test_zh.zh","w")
for l in zhs_test:
    f.write(l)
    f.write('\n')
f.close()
f = open("/rym/en-zh/train_zh.zh","w")
for l in zhs_train:
    f.write(l)
    f.write('\n')
f.close()

# 对保存的四个文件进行读取验证
file = open("/rym/en-zh/test_zh.zh")
tmp = []
while 1:
    line = file.readline()
    if not line:
        break
    count += 1
    tmp.append(line)
print(len(tmp))
tmp[:100]


###############测试集###############
file_zh = open("/rym/en-zh/test.zh")
file_en = open("/rym/en-zh/test.en")
zhs = []
ens = []
count = 0
N = 1000
while count < N:
    line_cn = file_zh.readline()
    line_en = file_en.readline()
    if len(line_cn) > 50:
```

```python
            zhs.append(line_cn)
            ens.append(line_en)
            count += 1
# 验证对齐
print(ens[998])
print(zhs[998])

df = pd.DataFrame('en':ens, 'zh':zhs)

## 去除句末换行符
for i in range(100):
        df.zh[i] = df.zh[i].replace('\n', '')
df['en'] = df.apply(lambda x:  x['en'].replace('\n', ''), axis=1)
df['zh'] = df.apply(lambda x:  x['zh'].replace('\n', ''), axis=1)
df_copy = df.copy()

## 去重
df = df.drop_duplicates()
df.to_csv("/rym/en-zh/seq_tmp_test.csv", index = False)

## 去除异常值
# 判断字符串是否包含中文
def is_contains_chinese(strs):
    for _char in strs:
        if '4e00'<= _char <= '9fa5':
            return True
    return False


def isvalid_zh(idx, string):
    if not isinstance(string, str):
        print("----不是字符串: ", idx, string)
        return False
    if len(string) < 2 or not is_contains_chinese(string):
        print("----不包含中文", idx, string)
        return False
#    if re.search(r"\W", string):
#        print("----有特殊字符", idx, string)
    return True
zhstrange_idx = [isvalid_zh(i, s) for i,s in enumerate(df['zh'])]
# 查看有多少条中文结果异常
```

```python
Counter(zhstrange_idx)

def isvalid_en(idx, string):
    if not isinstance(string, str):
        print("----不是字符串: ", idx, string)
        return False
#    if is_contains_chinese(string):
#        print("----包含中文: ", idx, string)
#        return False
    return True

enstrange_idx = [isvalid_en(i, s) for i,s in enumerate(df['en'])]
# 查看有多少条英文结果异常
Counter(enstrange_idx)
# 只要中文或英文异常即剔除
print(df.shape)
valid_index = [enstrange_idx[i] and zhstrange_idx[i] for i in range(df.shape[0])]
df = df[valid_index]
print(df.shape)
df.to_csv("/rym/en-zh/seq_clean_test.csv", index = False)

# 深度清理
ens = list(df['en'])
zhs = list(df['zh'])
del df

## 英文部分
replacement_patterns = [
    (r':', '&quot;'),
    (r'\"', '&apos;')
]
class RegexpReplacer(object):
    def __init__(self, patterns=replacement_patterns):
        self.patterns = [(re.compile(regex), repl) for (regex, repl) in patterns]
    def replace(self, text):
        s = text
        for (pattern, repl) in self.patterns:
            s = re.sub(pattern, repl, s)
        return s
```

```python
replacer = RegexpReplacer()
def en_process(ens):
    ens_new = []
    count = 0
    for en in ens:
        count += 1
        if(count%(10**5) == 0):
            print(count)
        en_tokens = nltk.word_tokenize(en)
        en_space = ' '.join(en_tokens)
        en = replacer.replace(en_space)
        ens_new.append(en)
    return(ens_new)

ens = en_process(ens)

f = open("/rym/en-zh/test2.en","w")
for l in ens:
    f.write(l)
    f.write('\n')
f.close()

## 中文
def zh_process(zhs):
    zhs_new = []
    count = 0
    for zh in zhs:
        count += 1
        if(count%(10**5) == 0):
            print(count)
        zh = zh.replace(" ", "")
        zh_tokens = list(jieba.cut(zh))
        zhs_new.append(' '.join(zh_tokens))
    return(zhs_new)

zhs = zh_process(zhs)

f = open("/rym/en-zh/test2.zh","w")
for l in zhs:
    f.write(l)
```

```
        f.write('\n')
f.close()

# 对保存的四个文件进行读取验证
file = open("/rym/en-zh/test2.zh")
tmp = []
while 1:
    line = file.readline()
    if not line:
        break
    count += 1
    tmp.append(line)
print(len(tmp))
tmp[:100]
```

6.1.3 BPE 分词

BPE 分词是一种数据压缩的方式，其对文本的处理过程可以简单地理解为一个单词的拆分再组合的过程。例如，对于 "loved" "loving" "loves" 这三个单词，BPE 通过训练，能够把它们拆分成 "lov" "ed" "ing" "es" 几部分，这样可以把词本身的意思和时态分开，有效地减少词表的数量。运行 BPE.md 文件中的 shell 命令，最终得到分词词表。分别选取中英文前 8 000 个词作为各自的词典。

6.2 数据描述统计

6.2.1 句子长度统计

首先统计原始句子长度，其中中文以字数计算，英文以空格分词计算，进行对数化处理后绘制频数分布直方图，如图 6.2(a) 所示。

中英文文本对数化句长后均呈右偏分布，英文句长整体多于中文句长。之后分析 BPE 分词结果，去除停用词，之后统计中英文文本分词结果每句的词数，中英文文本词数基本接近。分布具体结果如图 6.2(b) 所示。

6.2.2 词频统计

在对中英文文本分词后，统计每个词出现的次数，分别取中英文出现次数最多的前 20 个词，结果如图 6.3 所示。

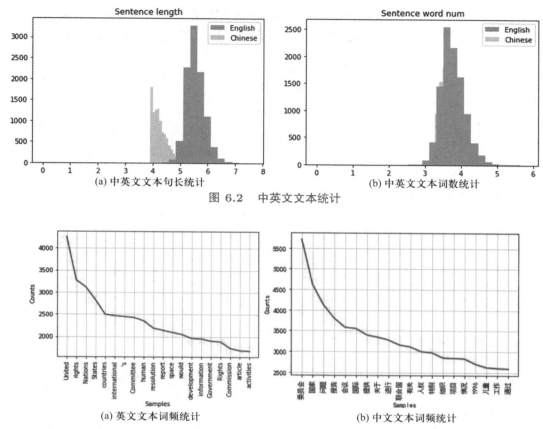

图 6.2 中英文文本统计

图 6.3 中英文文本词频统计

可以看到英文中最常出现的词是 "United" "rights" "Nations" 等，中文中最常出现 "委员会" "国家" "问题" 等. 为了进一步探究整体所有词汇出现的频次情况，绘制了中英文文本的词云图，如图 6.4 所示.

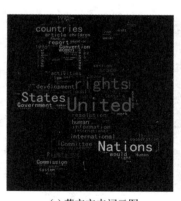
(a) 英文文本词云图

(b) 中文文本词云图

图 6.4 中英文文本词云图

可以看到该文本语料多与联合国委员会各项法案和会议等有关，这这进一步验证了语料来源有关于联合国法案文件，规范正式，适合用于模型训练. 以上部分见程序 desc_1.py.

```python
#句子描述统计
import pandas as pd
df = pd.read_csv('/rym/en-zh/seq_clean1.csv')
ens = df['en']
zhs = df['zh']

# 句子长度的描述统计
import matplotlib.pyplot as plt
import numpy as np
ens_sentlen = [len(en) for en in ens]
zhs_sentlen = [len(zh) for zh in zhs]
## 计算句子中单词个数
import nltk
import jieba

def get_ens_wordlen(ens):
    lens = []
    count = 0
    for en in ens:
        if count%(10**5) == 0:
            print(count)
        en_tokens = nltk.word_tokenize(en)
        en_tokens = [en_token for en_token in en_tokens if len(en_token) > 0]
        lens.append(len(en_tokens))
    return(lens)

ens_wordlen = get_ens_wordlen(ens)
def get_zhs_wordlen(zhs):
    lens = []
    count = 0
    for zh in zhs:
        count += 1
        if count%(10**5) == 0:
            print(count)
        zh_tokens = list(jieba.cut(zh))
        zh_tokens = [zh_token for zh_token in zh_tokens if len(zh_token) > 0]
        lens.append(len(zh_tokens))
    return(lens)
zhs_wordlen = get_zhs_wordlen(zhs)
```

```
## 存储结果
df_len = pd.DataFrame({'ens_sentlen':ens_sentlen,
                       'zhs_sentlen':zhs_sentlen,
                       'ens_wordlen':ens_wordlen,
                       'zhs_wordlen':zhs_wordlen,})

df_len.to_csv("/rym/en-zh/seq_lens.csv", index = False)

#绘图
ens_sentlen, zhs_sentlen, ens_wordlen, zhs_wordlen = df_len.iloc[:,0],
df_len.iloc[:,1], df_len.iloc[:,2], df_len.iloc[:,3]
plt.hist(np.log(ens_sentlen), bins=30, alpha=0.5)
plt.hist(np.log(zhs_sentlen),alpha=0.5, bins=30, color = "orange")
plt.legend(['English','Chinese'])
plt.title("Sentence length")

plt.boxplot([np.log(ens_sentlen/2), np.log(zhs_sentlen)],
            patch_artist=True, showfliers=False)
plt.hist(np.log(ens_wordlen), bins=30, alpha = 0.5)
plt.hist(np.log(zhs_wordlen), bins=30, alpha = 0.5, color = "orange")
plt.legend(['English','Chinese'])
plt.title("Sentence word num")

plt.boxplot([np.log(ens_wordlen), np.log(zhs_wordlen)], patch_artist=True,
showfliers=False)
```

6.2.3 词性统计

根据分词结果，对每个词进行词性标注，各取出现词性次数最多的前 20 个结果，可以看到中英文名词占比都是最多的，如图 6.5 所示，相关代码见 desc_2.py.

```
##词性统计
dat_en=[]
str_en = ""
i=0
for line in open("/rym/en-zh/train.en","r"):
    #en.write(line)
    str_en += line
    line=line[:-1]
    dat_en.append(line)
    i=i+1
```

```python
        if i>=10000:
            break
#en.close()

dat_zh=[]
str_zh = ""
i=0
for line in open("/rym/en-zh/train.zh","r"):
    #zh.write(line)
    str_zh += line
    line=line[:-1]
    dat_zh.append(line)
    i=i+1
    if i>=10000:
        break
#zh.close()

##去空、去重
def unde(one_list):
    '''
    使用列表推导的方式
    '''
    temp_list=[]
    for one in one_list:
        if one not in temp_list and one != "":
            temp_list.append(one)
        return temp_list

import string
a1=len(str_en)
b1=len(str_zh)
str_en=str_en.translate(str.maketrans('', '', string.punctuation))
str_zh=str_zh.translate(str.maketrans('', '', string.punctuation))
a2=len(str_en) #去除标点符号后
b2=len(str_zh)
print(a1-a2)
print(b1-b2)
import nltk
from collections import Counter
tokens = nltk.word_tokenize(str_en) #分词
tagged = nltk.pos_tag(tokens) #词性标注
```

```
tagged
ren = [x[1] for x in tagged]
dic_en=dict(Counter(ren))
dic_en
import jieba
import jieba.posseg as pseg
from collections import Counter
#用于词性标注
result = pseg.cut(str_zh)
rzh = []
#使用for循环把分出的词及其词性用/隔开,并添加,和空格
for w in result:
    rzh.append(w.flag)
dic_zh=dict(Counter(rzh))
dic_zh

import matplotlib.pyplot as plt
import numpy as np
# 创建一个点数为8 x 6 的窗口,并设置分辨率为80像素/每英寸
plt.figure(figsize=(10, 6), dpi=80)
# 再创建一个规格为1 x 1 的子图
# plt.subplot(1, 1, 1)
# 柱子总数
N = 10
# 包含每个柱子对应值的序列
dic_en1 = dict(sorted(dic_en.items(), key=lambda e:  e[1], reverse=True)[0:20])
values = list(dic_en1.values())
# 包含每个柱子下标的序列
index = list(dic_en1.keys())
# 柱子的宽度
width = 0.45
# 绘制柱状图,每根柱子的颜色为紫罗兰色
p2 = plt.bar(index, values, width, label="num", color="#87CEFA")
# 设置横轴标签
plt.xlabel('Part of speech')
# 设置纵轴标签
plt.ylabel('count')
# 添加标题
plt.title("Chinese")
# 添加纵横轴的刻度
plt.xticks(index)
# plt.yticks(np.arange(0, 10000, 10))
```

```python
# 添加图例
plt.legend(loc="upper right")
plt.show()


import matplotlib.pyplot as plt
import numpy as np
# 创建一个点数为8 x 6 的窗口，并设置分辨率为80像素/每英寸
plt.figure(figsize=(10, 6), dpi=80)
# 再创建一个规格为1 x 1 的子图
# plt.subplot(1, 1, 1)
# 柱子总数
N = 10
# 包含每个柱子对应值的序列
dic_zh1 = dict(sorted(dic_zh.items(), key=lambda e:  e[1], reverse=True)[0:20])
values = list(dic_zh1.values())
# 包含每个柱子下标的序列
index = list(dic_zh1.keys())
# 柱子的宽度
width = 0.45
# 绘制柱状图，每根柱子的颜色为紫罗兰色
p2 = plt.bar(index, values, width, label="num", color="#87CEFA")
# 设置横轴标签
plt.xlabel('Part of speech')
# 设置纵轴标签
plt.ylabel('count')
# 添加标题
plt.title("Chinese")
# 添加纵横轴的刻度
plt.xticks(index)
# plt.yticks(np.arange(0, 10000, 10))
# 添加图例
plt.legend(loc="upper right")
plt.show()
```

因为上述词性表达比较多、杂乱，所以我们对中英文进行了名词、动词和形副词的统计，观察两者表达的不同.

可以看出两者的占比都是名词> 动词> 形副词，明显不同的是名词、形副词的使用上英文要多于中文，而动词则是英文少于中文. 中英文在表达同一意思的情况下，用词侧重有所不同. 中文强调动作, 而英文更强调动作的发出者和承受者.

c_t 是由解码器第 $t-1$ 步的隐状态 h_{t-1} 与编码器中的每一个隐状态 $\bar{h}_s$ 加权计算得出的. 而第二种 Luong 注意力机制, 第 t 步的注意力 c_t 是由解码器第 t 步的隐状态 h_t 与编码器中的每一个隐状态 $\bar{h}_s$ 加权计算得出的, 后者逻辑上更自然, 但需要使用一层额外的网络来计算输出.

6.3.2 模型训练过程

在实际模型训练中, 为了加快随机梯度下降的收敛速度, 分别用了两个注意力机制权重归一化的版本进行训练, 最终的训练步数 num_train_steps 都是 20 000 步. num_layers=2, 使用两层的 RNN(LSTM); num_units=128, RNN(LSTM) 的隐层单元个数为 128 维; dropout=0.2, 使用的 Dropout 的概率为 0.2, 每个连接被保留的概率是 0.8. 这部分参数的设置可以在 s2s_attr/s2s_attr.md 中的前半段 shell 命令中设置并运行.

图 6.10(a)、(c) 是引入 Bahdanau 注意力机制的结果, 图 6.10(b)、(d) 是引入 Luong 注意力机制的结果, 可以看到训练集和测试集训练过程中复杂度都在逐步降低, 到最后趋向稳定, 复杂度越低, 建模效果越好. 使用 Luong 注意力机制相比之下复杂度更低一些, 测试集复

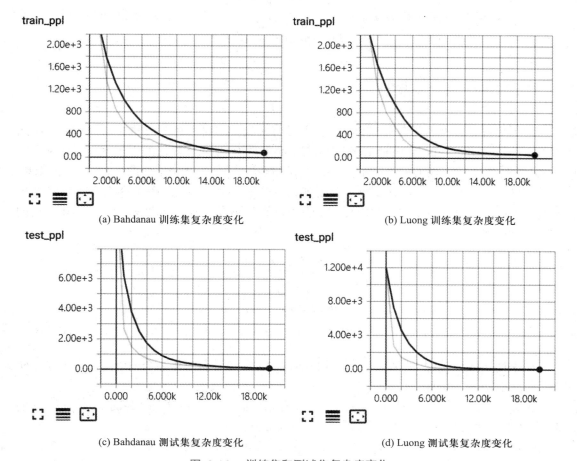

(a) Bahdanau 训练集复杂度变化　　　　(b) Luong 训练集复杂度变化

(c) Bahdanau 测试集复杂度变化　　　　(d) Luong 测试集复杂度变化

图 6.10　训练集和测试集复杂度变化

杂度降低到接近 0 的速度也比较快. 训练集损失函数都随着训练步数的增多, 整体在逐步下降, 如图 6.11 所示. 此处需运行 s2s_attr/s2s_attr.md 中的后半段 shell 命令.

本章使用的代码取自 github:tensorflow/nmt, 其中 nmt_master/nmt 文件夹包含了主要的模型代码. 使用 tensorboard 命令查看训练集、测试集相关指标变化 (使用 <tensorboard-port 22222-logdir/tmp/nmt_model/> 命令). (tensorboard 中图像往往有深浅两条线, 其中浅色的线代表了真实指标的曲线, 深色的线是经过平滑之后的曲线, 用于更好地展现曲线的趋势.)

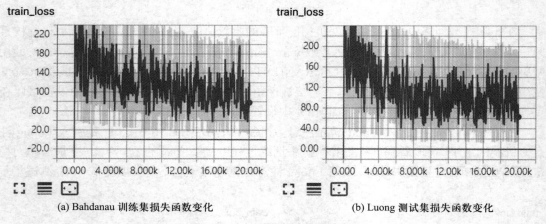

(a) Bahdanau 训练集损失函数变化　　(b) Luong 测试集损失函数变化

图 6.11　测试集损失函数变化

6.3.3　BLEU 值计算原理

为了更好地评测机器翻译的结果, 引入目前最流行的自动评测方法 —— IBM 提出的 BLEU (bilingual evaluation understudy) 算法. 简单来说, BLEU 算法的思想就是机器翻译的译文 (如图 6.12 中上面的一行所示) 越接近人工翻译 (如图 6.12 中下面的一行所示) 的结果, 它的翻译质量就越高. 所以评测算法就是如何定义机器翻译译文与参考译文之间的相似度. BLEU 采用一种 N-gram 的匹配规则, 原理比较简单, 就是考查译文和参考译文之间 n 组词的相似的占比.

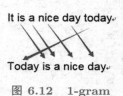

图 6.12　1-gram

如果用 1-gram 匹配的话. 可以看到机器译文一共 6 个词, 有 5 个词语都命中了参考译文, 那么它 1-gram 的匹配度为 5/6.

再以 3-gram 为例, 如图 6.13 所示, 可以看到机器译文一共可以分为四个 3-gram 的词组, 其中有两个可以命中参考译文, 那么它 3-gram 的匹配度为 2/4.

依次类推, 可以很容易实现一个程序来遍历计算 n-gram 的匹配度 ($n = 1, 2, \cdots, N$). 一般来说 1-gram 的结果代表了文中有多少个词被单独翻译出来了, 因此它反映的是这篇译文

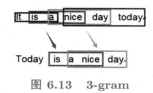

图 6.13 3-gram

的忠实度; 而当我们计算 2-gram 以上时, 更多时候结果反映的是译文的流畅度, 值越高文章的可读性就越好.

同时, 还需要考虑召回率. 将整个要处理的机器翻译的句子表示为 c_i, 如果参考译文不唯一, 有 m 个, 可以表示为 $s_i = s_{i_1}, s_{i_2}, \cdots, s_{i_m}$. n-gram 表示 n 个单词长度的词组集合, 对于第 k 个 n-gram, $h_k(c_i)$ 表示第 k 个 n-gram 在翻译译文 c_i 中出现的次数; $h_k(s_{i_j})$ 表示第 k 个 n-gram 在标准答案 s_{i_j} 中出现的次数.

综上所述, 各阶 n-gram 的精度都可以按照下面这个公式计算:

$$P_n = \frac{\sum_i \sum_k \min(h_k(c_i), \max_{j \in m} h_k(s_{i_j}))}{\sum_i \sum_k \min(h_k(c_i))}$$

然而 n-gram 的匹配度可能会随着句子长度变短而变好, 因此会存在这样一个问题: 一个翻译引擎只翻译出了句子中的一部分且翻译得比较准确, 那么它的匹配度依然会很高. 为了避免这种评分的偏向性, BLEU 在最后的评分结果中还引入了长度惩罚因子 (brevity penalty), 其计算方法如下:

$$BP = \begin{cases} 1, & \text{if } l_c > l_s \\ e^{1-\frac{l_s}{l_c}}, & \text{if } l_c \leqslant l_s \end{cases}$$

式中, l_c 为机器翻译译文的长度; l_s 为参考译文的有效长度, 当存在多个参考译文时, 选取和翻译译文最接近的长度. 当翻译译文长度大于参考译文的长度时, 惩罚系数为 1, 意味着不惩罚, 只有机器翻译译文长度小于参考答案才会计算惩罚因子.

由于各 n-gram 统计量的精度随着阶数的升高呈指数形式递减, 所以为了平衡各阶统计量的作用, 对其采用几何平均形式求平均值然后加权, 再乘以长度惩罚因子, 得到最后的评价公式:

$$BLEU = BP \times \exp\left(\sum_{n=1}^{N} W_n \log(P_n)\right)$$

BLEU 的原型系统采用的是均匀加权, 即 $W_n = 1/N$, N 的上限取值为 4, 即最多只统计 4-gram 的精度. BLEU 分数的计算见 BLEU.py.

```
#BLEU
from tqdm import tqdm
import numpy as np
## 数据读取
def get_data(path):
```

```
        file = open(path)
        tmp = []
        while 1:
            line = file.readline()
            if not line:
                break
            tmp.append(line)
        print(len(tmp))
        return(tmp)

out1 = get_data("nmt-master/tmp/nmt_attention_model/output_infer1.zh")
out2 = get_data("nmt-master/tmp/nmt_attention_model2/output_infer2.zh")
out3 = get_data("data/out_infer_trans.zh")
sdout = get_data("nmt-master/tmp/nmt_data/test2bpe.zh")

def process_res_sentence(sent):
    res = ''.join([s for s in sent.split(" ") if s != "<unk>"])
    res = [s for s in res][:-1]
    return(res)

# 分字，每一句话存为一个列表，每一个out都是存储列表的列表
out1 = [process_res_sentence(sent) for sent in tqdm(out1)]
out2 = [process_res_sentence(sent) for sent in tqdm(out2)]
out3 = [process_res_sentence(sent) for sent in tqdm(out3)]
sdout = [process_res_sentence(sent) for sent in tqdm(sdout)]

## 计算bleu 得分

from nltk.translate.bleu_score import corpus_bleu
sdout_corpus = [[s] for s in sdout]
res1_corpus = corpus_bleu(sdout_corpus, out1)
res2_corpus = corpus_bleu(sdout_corpus, out2)
res3_corpus = corpus_bleu(sdout_corpus, out3)
print(res1_corpus, res2_corpus, res3_corpus)
```

6.3.4 模型训练结果

按照上述 BLEU 值的计算原理，将测试集得到的结果进行分词处理再测算 BLEU 值，得到 normed_bahdanau 的 BLEU 值为 13.3，scaled_luong 的 BLEU 值为 15.6，采用 Luong 注意力机制训练模型结果较优，翻译效果较好。

6.4 Transformer 模型

2017 年谷歌公司在 "*Attention is All You Need*" 一文中提出了自注意力机制以及 Transformer 模型, 相对于传统的基于 LSTM 的 Seq2Seq 模型或者前文中实现的添加了注意力机制的模型, Transformer 通过自注意力机制可以让机器 "理解" 词句之间的关系, 对句子整个结构的把握和翻译的效果都会更好. 具体的模型结构和细节在 2.2.1 节已有介绍.

6.4.1 训练模型参数设置

在我们训练的 Transformer 模型中, 编码器和解码器都是由 6 个相对应的小单元堆叠而成, 而在 self-attention 层中, 是由 8 个 head 的自注意力机制构成. batchsize 为 128, 这是因为 Transformer 中需要训练的参数很多, 相比 Seq2Seq 就需要更多的训练轮数才可以得到较为满意的结果. 同样, 考虑到我们的训练集都是比较复杂的长句子, 将 learning_rate 设置为 0.003. 为了让参数充分收敛, 设置了 20 个 epochs, 在计算平台上大约耗时 3.5 小时. Transformer 模型代码见 transformer/train.py.

以上程序参数调整部分见 transformer/hparams.py, 其中 train_files 用于指定训练集和验证集的位置, training scheme 用于指定模型的 epochs、learning rate 等超参数, model 部分则是调整 Transformer 的网络结构.

其他程序和文件夹是运行 Transformer 模型所必需的代码文件. 本代码来自 github: Kyubyong/transformer 的实现.

6.4.2 训练结果

训练过程中的损失值和学习率如图 6.14 所示, 经过约为 20 个 epoch 的训练, 使用得到的模型翻译测试数据集 (1 000 条) 英文语句, 注意, 这里的测试集数据使用训练集得到的 BPE 词典进行分词, 对翻译结果计算 BLUE 值约为 18.4.

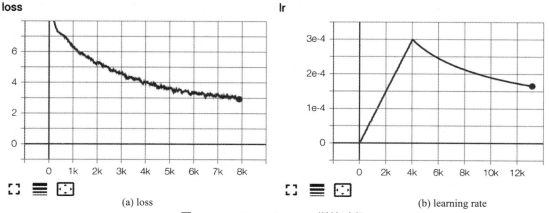

图 6.14　Transformer 训练过程

6.5 模型对比

对比三个模型在测试集上翻译结果的 BLEU 值, 可以看到加入 Bahdanau 的注意力机制的模型结果最差, Transformer 结果最好.

表 6.1 三个模型 BLEU 比较

	Seq2Seq+Bahdanau	Seq2Seq+Luong	Transformer
BLEU	13.3	14.6	18.4

由于 Transformer 模型的编码器端被设计成并行计算的模式, 其计算速度大幅提升, 无论速度还是效果都全面超越了以 Seq2Seq 为代表的 SOTA 模型, 这为后来的 Bert 埋下伏笔.

Seq2Seq 和 Transformer 模型在本案例中均可实现机器翻译的效果, 但是其内部机制并不相同, 单纯根据一个 BLEU 得分很难看到模型特点的差异. 读者可以自行选择样例, 使用不同的机器翻译模型进行翻译, 体会各个模型的翻译特点.

第 7 章 眼底图像分析示例

本案例基于眼底图像进行分析,演示一个基于图像数据进行分析的完整流程.医学影像分析是图像分析的一个重要应用领域,数据来源主要有 DR 影像、CT 影像、磁共振影像、病理影像、眼底影像等.基于医学影像开发自动诊断的产品是当前人工智能 (AI) 领域的热点,比较热门的领域是诊断肺癌、乳腺癌等.

由于一些专门的疾病诊断需要专业的医学知识做支撑,因此我们选择相对易于理解的眼底图像作为案例.现代研究发现很多疾病 (不仅是眼科疾病) 可以在眼底镜的图像中表现出来,适合利用大数据和图像技术来探索规律.由于眼底镜拍照非常容易,成本低,不具侵入性,很适合作为疾病筛查和初步诊断的依据.我们将在本案例中演示图像的分析和开发智能诊断工具的过程.

7.1 数据简介

本章所用数据来自北京大学 "智慧之眼" 国际眼底图像智能识别竞赛 (https://odir2019.grand-challenge.org),源自上工医信发布的经中国医学装备协会认证的包含 5 000 张眼底图像的标准数据集 "DDR",竞赛结束后,数据集 "DDR" 已面向社会公开.

数据集中的眼底图像由市场上的各种相机捕获,例如 Canon, Zeiss 和 Kowa, 因此产生了各种各样的图像分辨率.该数据集中病人的识别信息被移除.注释由经过培训的人类读者进行标记,并具有质量控制管理.他们将患者分为 8 个标签,包括正常 (N)、糖尿病 (D)、青光眼 (G)、白内障 (C)、AMD(A)、高血压 (H)、近视 (M) 和其他疾病/异常 (O).

很多疾病都会引起视网膜病变,也会体现到眼底镜的图像上,基于疾病标签可以训练分类模型,从而预测某张眼底镜图像属于哪一类疾病,实现智能诊断.在这个竞赛中,参与者必须提交所有测试数据集的 8 个类别的分类结果.对于每个类别,分类概率 (值从 0.0 到 1.0) 表示患者被诊断为具有相应类别的可能性.系统会自动对提交结果用三个指标进行评分:kappa 分数、F1 分数和 AUC 面积,阈值为 0.5,最终得分是上述指标的平均值.

图 7.1 显示了某位患者的眼底镜图像,分为左眼和右眼,可以看到清晰的眼底血管的形状,这是判断很多疾病的主要依据之一.此外,眼底图像中每个眼球的外侧中间部分有个反光的圆盘,称为视盘,也称视神经乳头,是视网膜上视觉纤维汇集穿出眼球的部位,是视神经的始端.在眼底图像的内侧有一个暗点,称为黄斑,是视锥细胞聚集的地方,也是视野的主要作用部位.

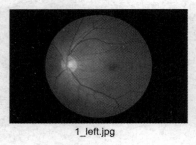 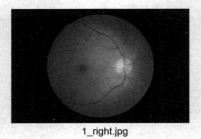

图 7.1 眼底数据示例

在实际的应用场景中，基于预测模型实现自动诊断的工具，可以帮助医生进行辅助诊断．但是需要注意的是，在医疗应用的场景，除了对模型的预测性要求比较高，还需要模型能解释，否则无法发现潜在的风险，在应用的时候会有问题．如果直接使用深度学习模型进行分类，通常很难解释，因此本案例使用深度学习模型进行血管分割，根据分割后的结果结合特征提取的方法建立统计模型，对特定疾病进行预测，从而兼顾模型的预测性和解释性.

为了简便起见，且演示一个完整的分析流程，我们挑选了 986 份包含正常人和糖尿病患者的样本，用来训练模型对糖尿病进行辅助诊断，这是一个两分类的问题，正负样本分别为糖尿病患者与正常人，数据包含眼底图像和疾病标签．DATA/imageinfo.xlsx 给出了数据的基本信息．其中第 3 列 class 中 N 表示正常样本，D 表示糖尿病样本．第 8 列 loc 表示样本来自左眼还是右眼．文件夹 image-origin 下是 986 张原始图片．数据的描述统计和建模将在 7.3 节进行介绍.

7.2 图像分割模型建立

7.2.1 数据预处理

对于眼底图像来说，最关键的信息是视网膜中的血管形态，受个体因素、光照环境、设备等影响，如果用完整的视网膜图像进行分析，会包含大量的噪声信息，因此首先要进行血管分割．"DDR" 数据集中包含了患者信息和疾病标签，但是并没有血管分割的标注，因此不能直接用来训练分割模型.

我们使用视网膜数据集 DRIVE(Digital Retinal Images for Vessel Extraction，其主页为 https://drive.grand-challenge.org) 进行训练，如图 7.2 所示．文件夹 DRIVE 保存了该数据集的训练与测试数据，image 文件夹下是原始数据，共有 20 张图片；1st_manual 文件夹下是对应的手工分割的结果，即打上标签的样本；mask 文件夹下是辅助背景图片信息.

原始图片分辨率较高，但对于眼底镜照片使用 584×565 左右的分辨率就可以满足分析的需求，还能够节约大量的计算资源．我们对原始照片进行处理，切除两旁的空白区域，并缩小分辨率，得到 584×565 分辨率的图像.

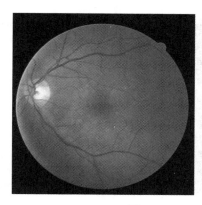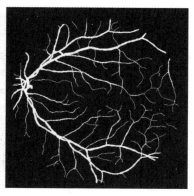

图 7.2 DRIVE 数据示例

由于照片的数量比较少,且单张照片的像素比较高,我们对图像进行增广处理,将其切割成很多份小照片,并调整成灰度模式,如图 7.3 所示.

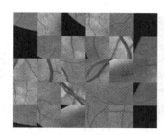

图 7.3 DRIVE 增广图像示例

最终将处理好的数据整合成 hdf5 文件格式,保存于 DRIVE_datasets_training_testing 文件夹下,以便于后续的模型训练. 以上数据预处理的代码见 prepare_datasets_DRIVE.py.

```
#=============================================================
#
# This prepare the hdf5 datasets of the DRIVE database
#
#=============================================================

import os
import h5py
import numpy as np
from PIL import Image

def write_hdf5(arr,outfile):
  with h5py.File(outfile,"w") as f:
      f.create_dataset("image", data=arr, dtype=arr.dtype)
```

```python
#------------Path of the images ---------------------------------------------
#train
original_imgs_train = "./DRIVE/training/images/"
groundTruth_imgs_train = "./DRIVE/training/1st_manual/"
borderMasks_imgs_train = "./DRIVE/training/mask/"
#test
original_imgs_test = "./DRIVE/test/images/"
groundTruth_imgs_test = "./DRIVE/test/1st_manual/"
borderMasks_imgs_test = "./DRIVE/test/mask/"
#----------------------------------------------------------------------------

Nimgs = 20
channels = 3
height = 584
width = 565
dataset_path = "./DRIVE_datasets_training_testing/"

def get_datasets(imgs_dir,groundTruth_dir,borderMasks_dir,train_test="null"):
    imgs = np.empty((Nimgs,height,width,channels))
    groundTruth = np.empty((Nimgs,height,width))
    border_masks = np.empty((Nimgs,height,width))
    for path, subdirs, files in os.walk(imgs_dir):  #list all files, directories in the path
        for i in range(len(files)):
            #original
            print("original image:  " +files[i])
            img = Image.open(imgs_dir+files[i])
            imgs[i] = np.asarray(img)
            #corresponding ground truth
            groundTruth_name = files[i][0:2] + "_manual1.gif"
            print("ground truth name:  " + groundTruth_name)
            g_truth = Image.open(groundTruth_dir + groundTruth_name)
            groundTruth[i] = np.asarray(g_truth)
            #corresponding border masks
            border_masks_name = ""
            if train_test=="train":
                border_masks_name = files[i][0:2] + "_training_mask.gif"
            elif train_test=="test":
                border_masks_name = files[i][0:2] + "_test_mask.gif"
            else:
```

```python
                print("specify if train or test!!")
                exit()
        print("border masks name:  " + border_masks_name)
        b_mask = Image.open(borderMasks_dir + border_masks_name)
        border_masks[i] = np.asarray(b_mask)

    print("imgs max:   " +str(np.max(imgs)))
    print("imgs min:   " +str(np.min(imgs)))
    assert(np.max(groundTruth)==255 and np.max(border_masks)==255)
    assert(np.min(groundTruth)==0 and np.min(border_masks)==0)
    print("ground truth and border masks are correctly withih pixel value range 0-255 (black-white)")
    #reshaping for my standard tensors
    imgs = np.transpose(imgs,(0,3,1,2))
    assert(imgs.shape == (Nimgs,channels,height,width))
    groundTruth = np.reshape(groundTruth,(Nimgs,1,height,width))
    border_masks = np.reshape(border_masks,(Nimgs,1,height,width))
    assert(groundTruth.shape == (Nimgs,1,height,width))
    assert(border_masks.shape == (Nimgs,1,height,width))
    return imgs, groundTruth, border_masks

if not os.path.exists(dataset_path):
    os.makedirs(dataset_path)
#getting the training datasets
imgs_train, groundTruth_train, border_masks_train = get_datasets(original_imgs_train,
groundTruth_imgs_train,borderMasks_imgs_train,"train")
print("saving train datasets")
write_hdf5(imgs_train, dataset_path + "DRIVE_dataset_imgs_train.hdf5")
write_hdf5(groundTruth_train, dataset_path + "DRIVE_dataset_groundTruth_train.hdf5")
write_hdf5(border_masks_train,dataset_path + "DRIVE_dataset_borderMasks_train.hdf5")

#getting the testing datasets
imgs_test, groundTruth_test, border_masks_test = get_datasets(original_imgs_test,
groundTruth_imgs_test,borderMasks_imgs_test,"test")
print("saving test datasets")
write_hdf5(imgs_test,dataset_path + "DRIVE_dataset_imgs_test.hdf5")
write_hdf5(groundTruth_test, dataset_path + "DRIVE_dataset_groundTruth_test.hdf5")
write_hdf5(border_masks_test,dataset_path + "DRIVE_dataset_borderMasks_test.hdf5")
```

7.2.2 模型训练

我们以原始图像的灰度图作为训练的特征集,把标注图像当作标签,将原始图像的每个像素对应到 "属于血管" 和 "不属于血管" 这两个类别,训练 U-Net 模型. U-Net 模型见 run_training.py 文件, lib 文件夹和 src 文件夹是 U-Net 模型的源代码.

```python
###################################################
#
# Script to launch the training
#
###################################################

import os, sys
import configparser

#config file to read from
config = configparser.RawConfigParser()
config.readfp(open(r'./configuration.txt'))
#===========================================
#name of the experiment
name_experiment = config.get('experiment name', 'name')
nohup = config.getboolean('training settings', 'nohup') #std output on log file?
# name_experiment = 'test'
# nohup = True


run_GPU = '' if sys.platform == 'win32'else ' THEANO_FLAGS=device=gpu,floatX=float32 '
#create a folder for the results
result_dir = name_experiment
print("\n1. Create directory for the results (if not already existing)")
if os.path.exists(result_dir):
    print("Dir already existing")
elif sys.platform=='win32':
    os.system('mkdir ' + result_dir)
else:
    os.system('mkdir -p ' +result_dir)

print("copy the configuration file in the results folder")
if sys.platform=='win32':
    os.system('copy configuration.txt .
'+name_experiment+'
'+name_experiment+'_configuration.txt')
```

```
else:
    os.system('cp configuration.txt ./'+name_experiment+'/'+name_experiment+'
    _configuration.txt')

# run the experiment
if nohup:
    print("\n2. Run the training on GPU with nohup")
    os.system(run_GPU +'nohup python -u ./src/retinaNN_training.py >
'+'./'+name_experiment+'/'+name_experiment+'_training.nohup')
else:
    print("\n2. Run the training on GPU (no nohup)")
    os.system(run_GPU +'python ./src/retinaNN_training.py'
)
#Prediction/testing is run with a different script
```

经过 20 次迭代后, 模型可以得到 0.952 的准确率, 精度召回的曲线下面积 (AUC) 为 0.896, F1 得分为 0.797. 基于该模型可以对采集到的其他眼底照片进行预测, 从而实现血管分割, 结果如图 7.4 所示.

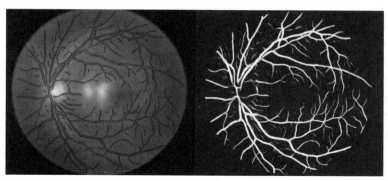

图 7.4　眼底血管分割结果

7.3　基于图像的智能诊断

7.3.1　图像分割结果

读者可以自己编写代码使用上述 U-Net 结果对本案例 DDR 数据中 986 张样本照片进行血管分割, 此处使用 R 软件的 medImageR 包. 这是一个专注医学图像的程序包, 在 Windows 操作系统中安装步骤如下:

(1) 在 Python 中安装 tensorflow==1.14, keras==2.2.4.

(2) 在 R 中安装 keras:

• devtools::install_version("keras", version="2.2.4").

(3) 在 R 语言中指定系统中 Python 的路径, 例如:
- library(keras);
- Sys.setenv(TENSORFLOW_PYTHON="<python path>");
- use_python("<python path>");
- install.packages("medImageR/medimageR_0.1-1.tar.gz",repo=NULL,type="source");
- library(medImage).

其中, medImageR 文件下对应安装该包所需要的源文件. 该包已经将 7.2.2 节模型训练的结果内置在其内部函数中, 可以直接调用相应的命令进行血管分割. 安装完成后, 可运行 1_segment.R 程序对本案例分析所用 DDR 数据 986 张图片进行血管分割, 结果保存在 /DATA/image_segment 文件夹下.

```
library(medimageR)

imgfiles <- list.files("image_origin", full.names = TRUE)
imgtif <- list.files("image_segment")
imgfiles <- imgfiles[!basename(imgfiles) %in% gsub("
.tiff", ".jpg", imgtif)]

for (i in 1:length(imgfiles)) {
    img1 <- readImage(imgfiles[i])
    img2 <- resizeRetina(img1)
    img3 <- outlineRetinaBV(img2)
    writeImage(img3, files = file.path("image_segment", gsub("
.jpg", ".tiff", basename(imgfiles[i]))), type = "tiff", quality = 100, bits.per.sample = 8L, compression = "LZW")
}
```

7.3.2 描述统计

在分割后的图像中, 首先提取血管的形态和颜色特征. 图 7.4 显示血管被分成了很多段, 可以通过图像技术进行特征提取, 可提取特征包括:
- 血管总面积 (area), 用来描述该眼球照片中血管的丰富程度.
- 血管平均宽度 (width), 用来描述血管的粗细情况.
- 血管细小分支的比例 (smallrate), 基于阈值提取细小分支, 计算比例.
- 血管的弯曲程度 (curved), 基于椭圆拟合计算每个分支的的弯曲程度.
- 血管细小分支的弯曲程度 (smallcurved), 可以衡量血管末端的扭曲程度.
- 血管的颜色特征, 包括颜色强度的均值 (bmean)、标准差 (bsd) 等.

结合之前介绍的图像纹理特征, 提取 Haralick 灰度共生矩阵的各项指标, 以及性别、年龄等人口统计学指标, 包含以下信息:
- 人口统计学指标: 性别 (sex) 和年龄 (age).

- Haralick 特征: 对比度 (con)、Haralick 相关性 (cor)、Haralick 方差 (var)、Haralick 同质度 (idm)、Haralick 和平均值 (sav)、Haralick 和方差 (sva)、Haralick 和熵 (sen)、Haralick 差分熵 (ent)、Haralick 差分方差 (dva)、Haralick 差分熵 (den)、Haralick 信息 f12 (f12)、Haralick 信息 f13 (f13)。

运行 2_analysis.R 的前半部分, 可以得到上述特征的数据结果. 以图像作为分析的粒度进行进一步描述统计分析 (见图 7.5), 可以看出男女比例、患病者与正常人的比例基本保持平衡, 正常人与患病者的年龄和性别都没有较大的差别, 可以看到这个数据集是一个相对均衡的数据集.

```
library(EBImage)
library(tmcn)
library(readxl)
library(dplyr)
library(e1071)
library(randomForest)
library(pROC)
library(rpart)
library(rpart.plot)
library(caret)
library(medimageR)
library(showtext)

infodf <- read_excel("imginfo.xlsx", sheet = "img")
varsdf <- read_excel("imginfo.xlsx", sheet = "vars")
infodf$sex <- factor(infodf$sex)
infodf$class <- factor(infodf$class)
levels(infodf$class) <- c("糖尿病","正常人")

# 提取图像特征
ftlist <- list()
for (i in 1:nrow(infodf)) {
    img0 <- readImage(file.path("image_origin", infodf$filename[i]))
    img0 <- resizeRetina(img0)
    img1 <- readImage(file.path("image_segment", gsub("\\.jpg", ".tiff",
infodf$filename[i])))
    img1 <- img1[3:(dim(img0)[1]+2), 1:dim(img0)[2]]

    cir1 <- matrix(0, dim(img1)[1], dim(img1)[2])
    cir2 <- drawCircle(cir1, x = dim(img1)[1] / 2, y = dim(img1)[2] / 2, radius =
min(dim(img1)[1:2]) / 2 - 5, col = 1, fill = TRUE)
    img2 <- img1
    img2@.Data[cir2 == 0] <- 0
```

```
    img3 <- img2
    img3@.Data[img3@.Data < 0.2] <- 0

    img4 = thresh(img3, 10, 10, 0.01)
    img4 = opening(img4, makeBrush(1, shape='disc'))
    img4 = bwlabel(img4)

    ft1 <- computeFeatures.shape(img4)
    ft2 <- computeFeatures.basic(img4, img0)
    ft3 <- computeFeatures.moment(img4, img0)
    ft4 <- computeFeatures.haralick(img4, img0)
    ft2 <- ft2[ft1[, "s.area"] > 5, ]
    ft3 <- ft3[ft1[, "s.area"] > 5, ]
    ft4 <- ft4[ft1[, "s.area"] > 5, ]
    ft1 <- ft1[ft1[, "s.area"] > 5, ]

    w0 <- as.numeric(ft1[, "s.area"]/ sum(ft1[, "s.area"]))
    ft4.tmp <- as.data.frame(ft4[, 1:13])
    for (j in 1:ncol(ft4.tmp)) {
        ft4.tmp[,j] <- ft4.tmp[,j] * w0
    }
    names(ft4.tmp) <- gsub("
..*?$", "", gsub("^.*?
.", "", names(ft4.tmp)))
    tmp.out <- data.frame(id = infodf$id[i], area = sum(ft1[, "s.area"]) / (pi *
(min(dim(img1))/2)^2),
        width = sum(ft1[, "s.area"])/sum(ft1[, "s.perimeter"]),
        smallrate = sum(ft1[ft1[, "s.area"] < 100, "s.area"]) / sum(ft1[, "s.area"]),
        curved = sum(ft3[, "m.eccentricity"] * w0),
        smallcurved = sum(ft3[ft1[, "s.area"] < 100, "m.eccentricity"] * w0[ft1[,
"s.area"] < 100]/sum(w0[ft1[, "s.area"] < 100])),
        bmean = sum(ft2[, "b.mean"] * w0), bsd = sum(ft2[, "b.sd"] * w0),
stringsAsFactors = FALSE)
    ftlist[[i]] <- cbind(tmp.out, as.data.frame(t(apply(ft4.tmp, 2, sum))))
}

ftdf <- infodf %>% left_join(do.call("rbind", ftlist))

font_add_google("Rock Salt", "rock")

showtext_auto()
```

```
# 描述统计
pdf(file = "anades01-1.pdf", width = 8, height = 5)
strvar <- varsdf$name[1]
par(mar = c(4,4,1,1))
layout(matrix(c(1, 2, 3, 3), 2, 2, byrow = TRUE))
boxplot(eval(parse(text = paste0(strvar, "~sex"))), data = ftdf, xlab = "性别")
plot(eval(parse(text = paste0(strvar, "~age"))), data = ftdf, xlab = "年龄")
lines(lowess(y = ftdf[[strvar]], x = ftdf$age), col = "red", lwd = 2, lty = 2)
boxplot(eval(parse(text = paste0(strvar, "~class"))), data = ftdf, cex.axis = 0.9,
xlab = "")
dev.off()

# Wilcox 检验
wilcox.test(x = ftdf[[strvar]][ftdf$class == "糖尿病"], y = ftdf[[strvar]][ftdf$class
== "正常人"])

# 逻辑斯蒂回归
mldf <- ftdf[, c(3, 4:5, 9:28)]
mldf$class <- factor(as.character(mldf$class))
mldf$sex <- factor(mldf$sex)
m1 <- glm(class~., data = mldf, family = "binomial")
prob1 <- predict(m1, newdata = mldf, type = "response")
roc1 <- roc(mldf$class, prob1)
p1 <- factor(c("糖尿病", "正常人")[as.numeric(prob1 > 0.5) + 1])
c1 <- confusionMatrix(p1, mldf$class, positive = "糖尿病")
c1

outdf1 <- as.data.frame(summary(m1)$coefficients)
for (i in 1:ncol(outdf1)) outdf1[[i]] <- round(outdf1[[i]], 3)
outdf1[["Signif."]] <- ""
for (i in 1:nrow(outdf1)) outdf1[["Signif."]][i] <- ifelse(outdf1[["Pr(>|z|)"]][i] <
0.01, "***",
        ifelse(outdf1[["Pr(>|z|)"]][i] < 0.05, "**", ifelse(outdf1[["Pr(>|z|)"]][i]
< 0.1, "*", "")))
rownames(outdf1) <- c("截距项", "年龄", "性别（女）", sapply(rownames(outdf1)[4:
nrow(outdf1)], FUN = function(X) varsdf$namecn[varsdf$name == X]))
outdf1

# 决策树
m2 <- rpart(class~., data = mldf, control = rpart.control(cp=0, minsplit =5, maxdepth
= 5))
```

```
prob2 <- as.numeric(predict(m2, mldf, type = "prob")[, "糖尿病"])
roc2 <- roc(mldf$class, prob2)
p2 <- predict(m2, mldf, type = "class")
c2 <- confusionMatrix(p2, mldf$class, positive = "糖尿病")
c2

pdf(file = "mdldt4-1.pdf", width = 8, height = 5)
rpart.plot(m2)
dev.off()

# ROC图
pdf(file = "mdlroc4-1.pdf", width = 11, height = 5)
par(mar = c(4,4,1,1), mfrow = c(1,2))
plot(roc1, print.auc = TRUE, print.thres = TRUE, main = "逻辑斯蒂回归")
plot(roc2, print.auc = TRUE, print.thres = TRUE, main = "决策树")
dev.off()

# 随机森林
m3 <- randomForest(class~., data = mldf, maxnodes = 50)
prob3 <- predict(m3, mldf, type = "prob")[, "糖尿病"]
roc3 <- roc(mldf$class, prob3)
p3 <- predict(m3, mldf, type = "response")
c3 <- confusionMatrix(p3, mldf$class, positive = "糖尿病")
c3
```

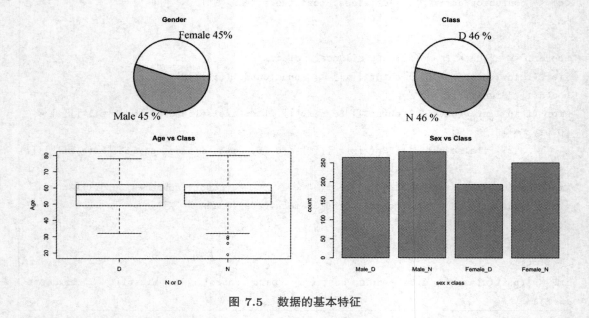

图 7.5 数据的基本特征

基于这些特征可以分析不同特征和疾病之间的关系，以血管总面积为例，图 7.6 显示了关于血管总面积指标的一些描述统计的结果, 从左上的图形可以看出不同性别的血管总面积不存在显著差异, 从右上的图形可以看出随着年龄的增长, 血管总面积存在明显的下降趋势. 下方的图形显示了不同疾病患者 (包括正常人) 的箱线图. 此处程序对应 2_analysis.R 中间部分.

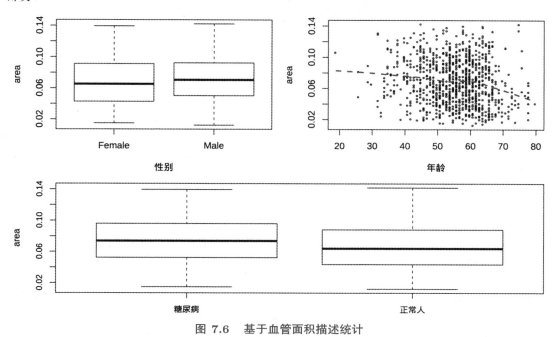

图 7.6 基于血管面积描述统计

7.3.3 诊断模型

基于之前对各个特征的分析, 可以发现这些特征和疾病之间存在一定的相关性, 因此我们把这些特征都作为自变量, 建立诊断模型. 对应程序 2_analysis.R 后半部分.

为了使得模型的解释性更好, 选择逻辑回归和决策树模型. 对于预测模型, 使用准确率 (Accuracy)、灵敏度 (Sensitivity)、特异性 (Specificity) 这三个常用指标进行评价. 在诊断的过程中, 灵敏度高表示不会遗漏患者, 特异性高表示不会误判健康人, 两者存在此消彼长的关系, 使用的时候需要根据实际需求进行权衡.

此外, 还会综合使用 ROC 曲线进行评估, 一般来说, 曲线的下面积 (AUC) 越大, 模型的效果也就越好.

首先尝试逻辑回归, 结果如表 7.1 所示, 其中 "Estimate" 列表示系数的估计值, "Std. Error" 表示标准误, "z value" 列表示对系数进行 Wald 检验的 Z 统计量的值, "Pr(>|z|)" 列表示系数的 P 值, "Signif." 表示显著性水平, "(*)" 表示显著性水平为 0.1, "(**)" 表示显著性水平为 0.05, "(***)" 表示显著性水平为 0.01.

结果显示, 血管总面积、血管细小分支的弯曲程度、血管颜色强度的标准差、Haralick 对比度、Haralick 方差、Haralick 和方差、Haralick 差分熵、Haralick 信息 f12 等指标对区分糖尿病患者与正常人的影像显著. 逻辑回归模型可以解释不同影响因素的重要程度, 因此在医学研究中使用很广泛.

表 7.1 逻辑回归结果

	Estimate	Std. Error	z value	Pr(> |z|)	Signif.
截距项	−22.235	17.379	−1.279	0.201	
年龄	−0.003	0.008	−0.402	0.687	
性别 (女)	−0.159	0.135	−1.179	0.238	
血管总面积	−14.709	4.463	−3.296	0.001	***
血管平均宽度	−0.35	0.436	−0.802	0.422	
血管细小分支的比例	−2.775	2.857	−0.971	0.331	
血管的弯曲程度	−0.189	0.773	−0.244	0.807	
血管细小分支的弯曲程度	23.196	7.284	3.184	0.001	***
血管颜色强度的均值	9.068	68.421	0.133	0.895	
血管颜色强度的标准差	82.539	42.133	1.959	0.05	*
Haralick 角二阶矩	−12.645	10.386	−1.217	0.223	
Haralick 对比度	4.422	1.936	2.284	0.022	**
Haralick 相关性	8.902	9.989	0.891	0.373	
Haralick 方差	−0.269	0.145	−1.851	0.064	*
Haralick 同质度	34.315	20.976	1.636	0.102	
Haralick 和平均值	−0.874	1.121	−0.78	0.435	
Haralick 和方差	0.008	0.003	3.164	0.002	***
Haralick 和熵	−22.201	25.464	−0.872	0.383	
Haralick 熵	23.391	23.386	1	0.317	
Haralick 差分熵	−58.382	23.931	−2.44	0.015	**
Haralick 信息 f12	−39.574	19.959	−1.983	0.047	**
Haralick 信息 f13	15.983	24.26	0.659	0.51	

基于训练集数据评估模型的预测效果, 可以得到模型的准确率为 0.592, 灵敏度为 0.473, 特异性为 0.696, 预测效果并不好, 可能原因包括模型不适合、特征有遗漏等, 通常可以尝试其他不同的模型.

继续使用决策树模型建模, 结果如图 7.7 所示, 可以得到一个诊断的树状规则, 每一个分叉表示基于一个变量进行评估, 符合条件后进入下一个节点, 最终得到的叶节点为分类的结果.

同样基于训练集数据评估训练误差, 可以得到模型的准确率为 0.64, 灵敏度为 0.72, 特异性为 0.571, 对比之前的逻辑回归模型效果略有提升.

然后计算两个模型的 ROC, 曲线如图 7.8 所示. 其中逻辑回归的下面积 AUC 的值为 0.634, 决策树模型的下面积 AUC 的值为 0.689, 可以认为决策树的效果更好.

通过逻辑回归可以解释不同的特征对判断糖尿病的影响程度, 通过决策树模型可以得到判断规则, 这两个模型的解释性都非常好, 也都可以用来预测, 不过在本例中, 直接使用的时候预测效果并不好.

在实际应用中, 还会把诸如逻辑回归和决策树这样的简单模型集成起来构成新的分类器, 称为集成模型, 预测性能通常会更好, 例如随机森林模型, 在本例中可以得到准确率为

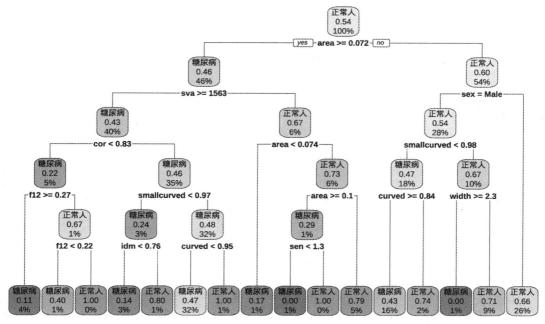

图 7.7 决策树模型结果

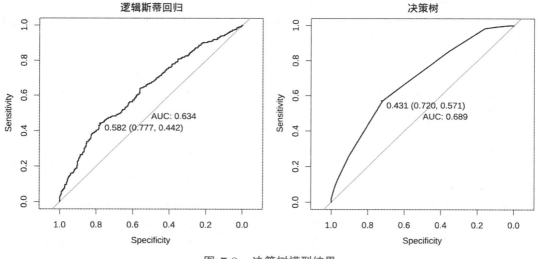

图 7.8 决策树模型结果

0.813 的结果.

选择合适的模型经过多轮训练后, 就可以得到一个最终的预测模型. 对于新的患者, 只需进行眼底拍照, 将数据输入到模型即可得到其罹患糖尿病的概率, 从而实现自动诊断.

参考文献

[1] Tukey J W. The Future of Data Analysis[J]. Annals of Mathematical Statistics, 1962, 33(1): 1-67.

[2] Box G E P. Science and Statistics. J Am Statist Assoc[J]. Journal of the American Statistical Association, 1976, 71(356): 791-799.

[3] Bin, Yu. Stability[J]. Bernoulli, 2013, 19(4): 1484-1500.

[4] Lecun Y, Bottou L, Bengio Y, et al. Gradient-Based Learning Applied to Document Recognition[J]. Proceedings of the IEEE, 1998, 86: 2278-2324.

[5] Krizhevsky A, Sutskever I, Hinton G. ImageNet Classification with Deep Convolutional Neural Networks[J]. Advances in Neural Information Processing Systems, 2012, 25(2).

[6] Simonyan K, Zisserman A. Very Deep Convolutional Networks for Large-Scale Image Recognition[J]. Computer Ence, 2014.

[7] He K, Zhang X, Ren S, et al. Deep Residual Learning for Image Recognition[C]//IEEE Conference on Computer Vision & Pattern Recognition. [S.l. : s.n.], 2016.

[8] Liu W, Anguelov D, Erhan D, et al. SSD: Single Shot MultiBox Detector[C]//European Conference on Computer Vision. [S.l. : s.n.], 2016.

[9] Girshick R, Donahue J, Darrell T, et al. Rich Feature Hierarchies for Accurate Object Detection and Semantic Segmentation[C]//Computer Vision and Pattern Recognition. [S.l. : s.n.], 2014.

[10] Long J, Shelhamer E, Darrell T. Fully Convolutional Networks for Semantic Segmentation[J]. IEEE Transactions on Pattern Analysis and Machine Intelligence, 2015, 39(4): 640-651.

[11] Ronneberger O, Fischer P, Brox T. U-Net: Convolutional Networks for Biomedical Image Segmentation[C]//LNCS: Medical Image Computing and Computer-Assisted Intervention (MICCAI): vol. 9351. [S.l.]: Springer, 2015: 234-241.

[12] Vaswani A, Shazeer N, Parmar N , et al. Attention Is All You Need[J]. arXiv, 2017.

图书在版编目（CIP）数据

数据科学实践／吕晓玲，李舰编著． -- 北京：中国人民大学出版社，2023．1
（数据科学与大数据技术丛书）
ISBN 978-7-300-31146-3

Ⅰ．①数… Ⅱ．①吕… ②李… Ⅲ．①数据处理 Ⅳ．①TP274

中国版本图书馆 CIP 数据核字（2022）第 198939 号

数据科学与大数据技术丛书
数据科学实践
吕晓玲　李　舰　编著
Shuju Kexue Shijian

出版发行	中国人民大学出版社			
社　　址	北京中关村大街 31 号		邮政编码	100080
电　　话	010－62511242（总编室）		010－62511770（质管部）	
	010－82501766（邮购部）		010－62514148（门市部）	
	010－62515195（发行公司）		010－62515275（盗版举报）	
网　　址	http://www.crup.com.cn			
经　　销	新华书店			
印　　刷	北京昌联印刷有限公司			
规　　格	185 mm×260 mm　16 开本		版　次	2023 年 1 月第 1 版
印　　张	12.25 插页 1		印　次	2023 年 1 月第 1 次印刷
字　　数	300 000		定　价	42.00 元

版权所有　侵权必究　印装差错　负责调换

中国人民大学出版社　管理分社

教师教学服务说明

中国人民大学出版社管理分社以出版经典、高品质的工商管理、统计、市场营销、人力资源管理、运营管理、物流管理、旅游管理等领域的各层次教材为宗旨。

为了更好地为一线教师服务，近年来管理分社着力建设了一批数字化、立体化的网络教学资源。教师可以通过以下方式获得免费下载教学资源的权限：

★ 在中国人民大学出版社网站 www.crup.com.cn 进行注册，注册后进入"会员中心"，在左侧点击"我的教师认证"，填写相关信息，提交后等待审核。我们将在一个工作日内为您开通相关资源的下载权限。

★ 如您急需教学资源或需要其他帮助，请加入教师 QQ 群或在工作时间与我们联络。

中国人民大学出版社　管理分社

- 教师 QQ 群：648333426（仅限教师加入）
- 联系电话：010-82501048，62515782，62515735
- 电子邮箱：glcbfs@crup.com.cn
- 通讯地址：北京市海淀区中关村大街甲 59 号文化大厦 1501 室（100872）